Telco Global Connect

TELCO GLOBAL CONNECT
Vol : 4

Author : Sadiq Malik

Copyright © 2016

ISBN-13: 978-1533087850

DEDICATION

" For Dr Jean Grey...my Phoenix Amazing "

This book is dedicated to the quest for a Digital Telco and explores an Agenda for Action. The Digital quest entails Smart Networks with enabling platforms, ecosystem integration, and customer analytics and new business models .The Digital Agenda entails building Smart Networks with enabling platforms, ecosystem integration, and customer analytics and new business models.

However any " digital ambition " must be scrutinised by asking some difficult questions and most of all formulating a strategic roadmap. Overcoming the organisational hurdles that sabotage such a difficult transformation journey is part of a new leadership paradigm.

In the future we may well see the emergence of new digital commodity exchanges that redefine existing businesses, by leveraging the global-local cloud infrastructure and dynamic connectivity to allow the facile exchange of digital goods or digitally-connected physical goods.

*" **We** become what we behold, we shape our tools and our tools shape us "*

Marshall McLuhan

PREFACE

Most experts concur that the pioneer and still champion of becoming a Digital Telco is Telefonica.That is not to say other Global Telcos have also digitised parts of their operations. However Telefonica put their money where there mouth is . To seize new opportunities in the Digital World Telefónica Digital was formed as a free standing business unit. Telefónica Digital had four main focus areas:

- Product Development and Innovation – the development of proprietary products and services through Telefónica R+D and other R&D operations
- Partnership & Venture Capital – Telefónica is committed to open innovation and where it cannot build products or services it will partner with, invest in or potentially acquire other companies large or small
- Innovative Digital Services – bringing to market, directly or through Telefónica operating businesses, new products and services across seven key segments – financial services, M2M, eHealth, advertising,video & media, security and cloud computing.
- New Business Areas – creating new business opportunities in areas such as over the top communications (TU), Big Data (Telefónica Dynamic Insights) and HTML5 (Open Web Devices).

Telefónica articulated its ambition and strategy to become a Digital Telco in a very simple but succinct manner. The aim is to empower customers to control, manage and enhance their daily life experiences through the use of digital services. These services are provided on an increasingly digital platform that will simplify service use and improve customers' overall experience. Telefonica realized that a digital Telco must build partnerships quickly to take advantage of new and potentially fleeting opportunities.

In 2015 Telefonica has outlined a five-year plan to become an 'Online Telco', under a new slogan 'We choose it all'. The plan is based on six foundations. Three relate to Telefonica's value proposition – outstanding connectivity, integrated or multiplay offerings, and a differentiated user experience. The other three are 'facilitators' of new digital services and strategies and consist of big data and innovation, end-to-end digitalisation, and capital allocation and simplification.

Telekom Italia (TI), in line with the evolution of the sector business models and market and technological trends, is also shifting towards a "Digital Telco & Platform Company" model based on innovative infrastructure and an excellent customer service,increasingly aimed at disseminating premium services and digital content within a customisable platform, accessible anywhere and on any device.

The usually conservative SingTEL surprised analyst with a sweeping reorganization in 2013. The company set aside $1.59 billion to fund a new division named Group Digital Life.The strategy is to invest in companies that would create new revenue streams in the online and digital world, addressing the conundrum now facing telecom companies: how to avoid life as a single-digit revenue growth utility in a world where the smartphone users can access Digital services from the Web.

Telcos are under constant pressure to optimize operational costs, gain agility and offer superior services to customers. In a wicked competitive environment , containing costs, streamlining operations, retaining customer loyalty, and maximizing the Average Margin Per User (AMPU) becomes a business imperative. Short product life cycles and overheated marketing are overwhelming the operators, which resort to ad hoc solutions that appear to offer customers what they want, but in fact mask additional costs. These costs, however, may not show up until further down the service delivery chain in other areas of the business, where their root cause may be understood but cannot be addressed across functional boundaries.

Complexity is a fact of life for telecom operators, but it is also a cost driver. Legacy systems are maintained alongside next-generation networks .The complexity that has overtaken the telecom business has resulted in organizations with technology frameworks, tariff structures, and product catalogues that if plotted on a chart would resemble a Jackson Pollock painting. One European operator found that it was offering 20,000 different tariffs to 15 million customers in one country; after it streamlined its processes to respond to real customer needs, the number of tariffs was reduced to 8,000 !! Analysis by the BCG shows that Telecoms is one of the most inefficient industries with over 40 % of its cost base gobbled up by waste in various telecoms processes.

So what do we do to root out waste and transform ?? To start with use the TM Forums's Business Process Framework (eTOM) : a widely deployed and accepted model and framework for business processes in the Information, Communications, and Entertainment industries. As a key part of TM Forum's Frameworx, the Business Process Framework represents the whole of a Service Provider's enterprise environment in a hierarchy of process elements that capture process detail at various levels.The Business Process Framework (eTOM) describes and analyzes different levels of enterprise processes according to their significance and priority for the business . For CSP's , the Business Process Framework serves as the blueprint for process direction. Here are some case studies to vindicate e TOM's effectiveness.

Qwest wanted to transform its service delivery to shorten the time-to-market for new products, including cloud services, reduce its operating costs, and have visibility and traceability from products to services to resources. It was also determined to reduce individual service component redundancy and enforce Qwest's high standards for the overall customer experience.To reduce investment risk and prove the viability of what it wanted to achieve, the operator and its partners turned to TM Forum's Frameworx and Catalyst Program before it embarked on the transformation. Within a year of the deployment Qwest saw a 4 percent increase in revenue, a 5 percent cost reduction, a 25 percent improvement in new product deployment cycle times, and a decrease in unique provisioning and assurance job steps.

Magyar Telekom's project to convert a legacy provisioning system into a single platform successfully enables the provisioning and activation of multiple product lines. The implementation relied heavily on TM Forum Frameworx and is delivering many benefits. They include cutting service activation by 20 percent and increasing the ratio of successful automated activations by 30 percent. Time-to-market for services was reduced by up to 20 percent, while the time needed to integrate new network management systems fell by 30 percent. When manual interventions are needed, they take 70 percent less time. The deployment of a zero-touch home gateway has lessened field force activity by 30 percent. New and existing services are being migrated to a new platform, and CRM will be enhanced to support trouble ticketing and the management of service level agreements.

Concurrently with e TOM an approach called Lean Six Sigma (used so succefully in corporates from the Fortune 500 financial , manufacturing and service industries) can be implemented to cut waste and inefficiency in Telco processes. Lean Six Sigma is a managerial concept combining Lean and Six Sigma that results in the elimination of the eight kinds of wastes / muda (classified as Defects, Overproduction, Waiting, Non-Utilized Talent, Transportation, Inventory, Motion, Extra-Processing) and provision of goods and service at a rate of 3.4 defects per million opportunities (DPMO). Lean Six Sigma utilises the DMAIC phases similar to that of Six Sigma. DMAIC (an abbreviation for Define, Measure, Analyze, Improve and Control) refers to a data-driven improvement cycle used for improving, optimizing and stabilizing business processes and designs.

There are 4 overarching strategies in the endeavour to create a LEAN MEAN Organisation.The categories can be seen as structural, transformational, changes with high complexity. Pursuing any of these should not be seen as a replacement to the first strategy of continuous improvement – there is always something more that can be done to improve the efficiency within the business as it is today.

- Improve cost efficiency and productivity through automation, centralisation, market differentiation and reengineering of work processes (including partnering)
- Realise national economy of scale by mergers & acquisitions with competing operators (including network sharing)
- Achieve international economy of scale by implementation of cross-border working processes
- Leverage national economy of scope by integrating fixed, broadband, TV or mobile businesses

Today, every aspect of your telco's operations needs to be measured against the touchstone of COST EFFICIENCY to ensure it brings profits. Whether it is investment in Transformative IP programs, in merging and acquiring enterprises, or in outpacing competition; eliminating redundancies and optimizing processes is essential. Getting rid of people (to cut costs) hardly requires imagination unless senior execs have overloaded the Company with relatives , buddies and PA's who double up as girlfriends !! Using both e TOM and Lean Six Sigma Telcos can cut costs INTELLIGENTLY in a variety of areas as identified by the gurus at AT Kearney :

Network, marketing, and IT : These three areas have the most potential for optimizing operational and capital expenditures, typically by reducing complexity.

Supply chain and procurement : Some Global Telcos aspire rapid international growth—often through acquisitions presents plenty of opportunities to improve supply chain and procurement capabilities. By standardizing purchasing requirements and internal technical specifications, consolidating volumes, and optimizing deals with suppliers, operators can cut costs without affecting core operations.

Back office : Consolidating back-office functions such as HR and finance, potentially by establishing central or regional shared services, can increase efficiency

Information technology : Centralizing IT services and standardizing or consolidating applications and hardware can substantially reduce costs and often improve service.

Infrastructure sharing : Sharing infrastructure among operators is another way to optimize costs and leverage economies of scale. For example, Bharti, Millicom, and Vodafone (Spain, Germany, U.K., India, and Ireland) have shared networks with other operators. In Sweden, 3 and Telenor's joint venture, 3GIS, covers around 70 percent of its network with shared infrastructure.

Outsourcing: Outsourcing non-core activities, such as fleet services , advertising and facility management, can improve efficiency and allow more management focus on customers. Newer outsourcing models include managed capacity, where an outsourcer is paid on a variable utilization or capacity basis. These models, besides increasing efficiency, reduce risk, and limit financing needs while fundamentally shifting the focus from operations to customer experience and partnership management.

Energy efficiency : Energy efficiency can cut costs while reducing environmental impact. France Telecom-Orange, for example, is aiming to reduce energy consumption by 15 percent between 2006 and 2020. By the end of 2010, the group had fitted more than 8,000 network sites with optimized ventilation systems, cut energy consumption at data centers, and installed solar-powered base stations (mainly in Africa and the Middle East).

Telenor, for example, reduced its software licensing costs by 34 percent by replacing local licensing agreements with global deals.Telcos will need to use the full scale of their groups to create synergies, reduce external spending, and benefit from solid supplier relationships, which can bring earlier access to new handsets and network equipment .Bharti Airtel's so-called "Minutes Factory" has enabled it to target millions of pre-paid customers that would have been too costly to serve using the conventional subscriber-led model. The factory's key elements include outsourced network equipment, which enables fixed costs to convert to variable costs. Bharti's partnerships enable it to add network and IT capacity quickly and efficiently, as needed.

We know that taming complexity and streamlining operations can reduce operational costs by a third and provide customers with better service. The up-front savings achievable in the short term—six to 12 months—will cover the costs of the initial assessment that identifies how and where to implement a Lean transformation using eTOM and Six Sigma methodologies.The good news for telecom companies faced with stalled revenue growth is that there are ways to significantly reduce expenditures. Telecom carriers will have to lower their operating expenses for traditional telecom services to maximize free cash flow, which can be invested in nontraditional services. Telcos must focus on operating efficiency when offering a suite of non traditional services in the 4G data world services, as there are no "killer" applications.

Operators that do not undergo a lean transformation, however, will find themselves unable to compete. The decision to adopt a lean and mean approach needs to be made before you become extinct !!

In MEA , fixed and mobile Telcos have not fully realized the cost-reduction potential provided by lean tools and techniques, which not only can generate savings of from 10 to 15 percent on the addressable cost base, but also simultaneously improve overall operational quality levels.This process should start with a diagnostic phase that covers network planning and implementation, operations, and management infrastructure. There are also a broad range of new OPEX saving possibilities which can be leveraged through a new generation of technologies, e. g. in the area of software defined radio networks (SDR) and self organized networks (SON).

The network operations centers of many telcos face a variety of challenges, including having to deal with technology silos, unclear ownership of network issues, lack of institutional memory that forces teams to "reinvent the wheel" time and again, and others. Given the breadth of opportunities available, operators can often capture reductions of 15 to 35 percent in NOC-related costs. Potential actions include developing a clean-sheet NOC redesign, integrating NOC services on an end-to-end basis, and instilling a problem-solving, high performance mindset within the center. By introducing optimized governance models, best-practice vendor relationship management techniques, and better negotiation and deal strategies, operators that revisit mobile outsourcing typically identify the potential for an additional 5 to 10 percent in cost reduction, representing 2 to 3 percent of total costs.

Mobile operators can make use of the rich variety of customer data they have on hand to improve their network quality and target investments on a site-by-site basis. Taking this type of highly granular review of network performance metrics, site utilization, and commercial performance will enable leaders to pinpoint spending requirements.By using techniques such as network caching and CDN, operators can reduce one-on-one network downloading and hence, network load and form partnerships with broadcasters to share investments and build a large-scale, secure, single network infrastructure.

Telco Gurus believe that even that fixed-line infrastructure players will outsource network infrastructure and operation to contractors in order to optimize operating and capital expenditures (opex and capex). Making this outsourcing a success requires companies to explicitly split roles and responsibilities with the chosen contractors, establish clear reporting and interface models, and prepare, negotiate, and execute specific contracts and service level agreements. Telecoms players can employ proprietary analyses and techniques to improve the amount of value their products deliver to customers, while at the same time, creating cost-efficient designs and calculating target costs.

While personnel wages and benefits represent a major network operating cost, other high potential areas for cost cutting include site rental and energy costs. As a consequence, some operators are aggressively pursuing the renegotiation of rental contracts with an eye toward moving or eliminating those sites with the most expensive rental contracts. Considering network optimization, some operators are exploring base transceiver station (BTS) "hotels." These BTS hotels group the electronics from a number of base stations for antennae up to 20 km away.

The priorities of telecom CIOs are related to business applications, BI ,analytics, customer service support systems, centralizing the billing systems and the convergence of Internet protocol. While CIOs work to deliver applications and innovation, they are also asked to be as cost effective as possible. Looking ahead, IT leaders from communications service providers need to be aware of the shifts brought by new technologies – and they will be requested to provide insights on how carriers should provide strategic plans and new services to be launched based on them. This may involve changes to the infrastructure layer to fit new demands.

The new, digitally connected world is driving transformation, bringing with it new players, advanced applications, broadband services and higher QoS demands. Seizing this transformation by making your business and technology evolution work together is the key to profiting from market changes.BT has undergone a major transformation and continues to change. What was a traditional networks-focused, telco R&D organisation is now a 'softco', centred on developing software and using software development methodologies and practices. At the same time, networks and computing are quickly converging into what we know as cloud computing. Not surprisingly, BT's continuing transformation is now addressing the cloud services market.

A basic cost reduction mechanism and culture across all staff must be in place (e. g. personal target setting, cost transparency, etc.). The challenge is to embed an organisational discipline that will constantly challenge the existing cost basis. The benefits of creating a performance-driven culture within Telcos come from its capacity to amplify subsequent improvement initiatives – in effect, supercharging them. However, as with most transformational approaches, "getting there" requires strong, visible commitment from company leaders, solid organizational planning and training, and communication clarity. One thing for sure : solving the cost/efficiency puzzle requires a wholistic mult facet approach that will target the right levers to optimise cost even as capex is injected into building high speed IP based broadband networks.

Telecommunications companies are facing difficult times these days: their products are commoditised, competition is fierce and pressure on margins is high. Some operators with foresight are trying to set themselves apart through better service. Services, by being less visible and more labor dependent, are much more difficult to imitate, thus becoming a sustainable source of competitive advantage. Reaching out to and understanding the needs of customers, both current ones and those who may consider shifting from competitors in such turbulent and competitive era, is an important element of the service strategy. As such Telcos would do well to learn from other industry best practices to gain a competitive edge in and expand beyond their core markets.

Other industries, such as computer or telecom hardware, have shown that business development based on new services can be a successful road to new growth. In 1993, CEO Louis Gerstner initiated a transformation at IBM. The transformation involved a change from a hardware and software business to a solutions and services business and from a regionally aligned organization to a global organization. IBM committed itself to business and cultural change , invested in talent and the right financial and IT systems to support them and placed strategic bets on IT as a utility service and hosted storage. The transformation created opportunities for cost savings by encouraging development and use of enterprise-wide technology platforms.

Fortunately with the ever evolving technology and more complex products, incumbent operators have a powerful asset to leverage: their technical field service organization and capabilities. In order to capitalise on further strategic growth opportunities, Telcos should consider an option of developing a dedicated service organization which has control of its entire value chain, primarily focusing on multiproduct communications services for mass segments. This means a dedicated unit, focusing on developing the service business, having full control over its entire value chain, freedom and leeway to develop its business, full management attention and support to execute its mandate.

The primary focus of the new service organizations is service innovation excellence and ability to scale customer solutions for rapid growth across well defined customer segments based on their real communications needs. Customer front-end responsibility (marketing and sales) should be placed into new service Org. The consolidation of the service offering under a single division is normally accompanied by a strong initiative to improve the efficiency, quality and delivery time of the services provided, and the creation of additional services to supplement the basic offering. The consolidation of services also comes with the development of a monitoring system to assess the effectiveness and efficiency of the service delivery.

Transitioning from product manufacturer into service provider constitutes some managerial challenges. Services require new organizational principles, structures and processes. Not only are new capabilities, metrics and incentives needed, but also the emphasis of the business model changes from transaction- to relationship-based. Be warned that research has shown it takes a serious effort by senior management to build the structures, capabilities, processes and systems to seize the service opportunities. Successful service companies do not start from scratch – they are built up on the basis of existing units and businesses with the best suited set of service assets and capabilities such as customer knowledge ; service development, standardization and roll-out capabilities for complete service delivery.

A focus on forward-looking IT investments (funded by reductions in maintenance costs for today's systems) will be essential to support the service organisation. Social media collaboration platforms support all service operations: enterprise case management, call centers, customer portals, websites, and integration with social media channels. A knowledge base provides answers to your agents and your customers through all your channels, increasing deflection rates and reducing time spent per case, keeping your customers happy and loyal. Cloud based CRM platforms can support the customer service team to improve the way they managed everyday customer interactions.

Communications service providers have a number of attributes that give them a potential marketplace advantage: an extensive customer base, distribution muscle and knowledge of customer preferences through CRM and billing systems. The opportunity is to become an integrated digital services provider across platforms and mobile devices—convincing customers that a communications service provider can effectively serve as the hub to meet their communication and entertainment needs. According to an Accenture survey, the areas that show particular promise include cloud services and location-based offers.

Cloud Customer Portals give customers a true online service experience, ensuring they have the flexibility to manage their interactions with their Telco entirely online if they chose to do so, and enabling customer service questions to be managed, just like an order coming in from a field sales team person. Customers can create orders online for new and replacement products including, phones, accessories, and SIM cards, and then track the status of the order through to shipment. When Sprint acquired Nextel, over 5,000 employees in over 1,100 retail stores and 800 dealer locations got busy collaborating on customer retention and churn avoidance. Tied together via an employee social network, disparate teams across both organizations focused their efforts to retain customers and build new loyalty programs.

Delivering good services as part of a core product offering does not suffice as the sole differentiator in highly competitive telecommunications markets. Investments in new radio access technology bring along radically new network economics leaving mobile operators with the quest to gear their network investments towards a cost optimal access, backhaul and core portfolio. It is critical to cut spending on low-value activities, and redeploy it to investments that generate growth, margins and true differentiation. Being able to accurately identify where value is generated at all levels of the organization – from divisions to specific products or offerings to particular customers – is a critical managerial competence.

Customer ownership and distribution power give communications service providers a strong foundation on which to build to meet consumers' ongoing communication and entertainment needs. Providers have an opportunity to improve their return on investment by monetizing better connectivity. They can also extend their partnerships across the digital ecosystem to provide a seamless customer experience. This will require deep insight into subscriber behaviors, new forms of collaboration within the industry, new capabilities within the organization and an ability to constantly innovate to keep pace with today's demanding consumers.

Perhaps one of the most successful new age " experience " players to date is SK Planet, which was set up in 2011 by SK Telecom, Korea's largest wireless operator, to offer multiple add-on experiences for both retail and business subscribers. They include MelOn, already Korea's largest music portal, with 17 million subscribers, has also been launched in Indonesia; 11st provides an e-commerce platform with related advertising and marketing intelligence services.It is now the country's second-largest e-commerce platform and largest player in mobile commerce; "T ad" is a mobile ad platform that enables personalized ads on mobile apps running on smartphones and tablets; "T map," a GPS-based navigation service platform with more than 10 million subscribers, also offers location-based services to businesses.

By consolidating a wide range of services under one roof, on top of its successful core wireless broadband business, SK Planet now offers perhaps one of the most compelling customer experiences of any operator worldwide.To provide customers with exactly the right products and services based on their actions and requests, " experience " players such as SK have to become fully responsive to the correct interpretations of their customers' behavior, often in real time. The ability, for instance, to offer access to medical information services could follow evidence of increased interest in healthcare. So investing in data analytics capabilities to respond to customer data is a must. "Big data" offers much promise in this area, but it will require considerable investment.

According to Booz " experience play " Telcos de-emphasize their network activities and in some cases even carving out their entire access network infrastructure—both passive and active. They can share these costly assets with competitors via network-sharing agreements, and then differentiate themselves through innovative products and services. In contrast, some recent mobile and fixed entrants in Europe and the Middle East have focused on deploying their own network infrastructure in order to become connectivity or platform players. Consequently, they have minimized their investments in customer-facing infrastructure, relying on an online presence for sales and deploying only flagship stores to serve as their bricks-and-mortar channel.

Telecom operators should look for opportunities for growth by both assessing their markets (the market-back view) and evaluating their own current strengths (the capabilities-forward view). Both points of view are critical; if the two aren't considered together, the result will likely be a capabilities system poorly aligned with market opportunities. The capabilities forward view seeks to find distinctive internal capabilities that can be leveraged in any number of ways: to grow into adjacent markets, to build innovative new services, or to increase network speed and capacity.The market-back view turns outward for market opportunities that might arise from new technologies or from opportunities that competitors might be overlooking or are not coherently pursuing.The goal is to become coherent: to strike a balance so that the right product and service portfolio naturally thrives within a capabilities system consciously chosen implemented

The transformation program must set a clear vision for the entire organization as it improves operational excellence and moves towards the defined target architecture. Create a high-level transformation checklist that covers a number of tasks starting with problem acknowledgment, followed by executive buy-in and budget approval. The Transformation agenda is based on the following pillars :

1. Analysing the current and future technology investments from business and technical viewpoints. In addition review the key customer drivers and applications that generate fast ROI based on understanding the needs of target markets. Conduct a network inventory to identify and retire and redundant network elements

2. Investigate how to incorporate Web 2 / Telco 2 paradigms into the creation of your product portfolio. Web2.0 is the new generation of web services, characterized as more open, flexible and participatory in terms of creating content, applications and collaborative alliances.

3. Strategise capabilities to overcome OTT net players to make money from higher value added services by implementing " smart pipe "design. We must assess every aspect of the network from its underlying hardware and systems to its configuration, capacity, traffic flow, and survivability.

4. Perform a thorough evaluation of the operations, identify the gaps between the current methods and the future vision, define new job functions and processes, prepare a road map for transformation, and facilitate the migration process.

5. Streamline the architecture with the judicious use of web services and services-oriented architecture (SOA). In addition to streamlining the network management systems environment, platforms and tools that enable service and customer management need to be introduced.

6. People and human resource skills must be upgraded to meet the needs of the new organization structure and new IP based technologies .Nimble, efficient operations rely on modern business processes, management practices, and human resources

7. Have a clear view on the risks and formulating mitigation strategies. Risk assessment must extend beyond the usual financial and regulatory risks to consider the wider environment in which the organization operates and the full extent of its operations, now and into the future. A failure to shift the business model from minutes to bytes , misunderstanding the changing customer mindset and insufficient insight into latent data assets are such risks.

SingTel is Asia's leading communications group with operations and investments around the world .With significant operations in Singapore and Australia (through wholly-owned subsidiary SingTel Optus), the Group provides a comprehensive portfolio of services that includes voice and data solutions over fixed, wireless and Internet platforms, as well as infocomm technology and pay TV. The Group has presence in Asia and Africa with 473 million mobile customers in 26 countries, including Bangladesh, India, Indonesia, the Philippines and Thailand.

SingTel's domestic operation, is a typical full-service converged carrier. Currently, their top priority is the deployment of 4G wireless and FTTH in the fixed sphere. SingTel is cooperating in the Singaporean government's national broadband network project, which aims to provide 1Gbps service and to offer full openness at Layer 2 and also at Layer 0 (i.e. to the ducts).SingTel are also a very significant force in carrier services, wholesale, and IP transit. SingTel is also aiming to double the size of its satellite business, with two satellite launches scheduled within the next two years via JV's.

Recently the operator decided to restructure its business into three units : group consumer, group digital life and group ICT-in order to sustain growth, competitiveness, and innovation. With the reorganization, SingTel plan to reinvent its core carriage business, create and drive new growth platforms that leverage and strengthen the core, and turbo-charge its regional capabilities in ICT services. They broke new ground with the introduction of PowerON Compute Service. This state-of-the-art cloud solution provides enterprises with the business agility and cost effectiveness of public clouds without compromising on portability, compatibility, security and control demanded by enterprise IT organisations.

SingTel is banking on acquisitions of smaller companies to help drive growth in its business as revenue slows from mature markets like Singapore and Australia. SingTel plans to spend US$1.6 billion in three years to acquire companies specializing in digital advertising, content and entertainment. That $1.6 billion investment will be spread over the next three years, and will be largely ploughed into strategic acquisitions in the online space that can tie in with SingTel's phone services across the region. SingTel spent US$400 million in acquiring advertising, entertainment and digital commerce firms in the last fiscal year, including the US$321 billion it paid for Amobee, a U.S.-based mobile advertising company. By leveraging their unique assets and Amobee, they will be able to realise the full potential of mobile marketing as a platform to change the way brands communicate with their customers.

The company is also using surplus cash to step up dividend payments to keep investors happy even as it continues to invest in its core telecommunications business. SingTel's dividend payout ratio ranges from 55 per cent to 70 per cent of underlying net profit. The Group will continue to review at least on a three-year basis its cash needs for operations and growth, with a view to returning surplus cash to shareholders. This is consistent with the Group's commitment to an optimal capital structure and investment grade credit ratings, while maintaining financial flexibility.

Companies like SingTel go beyond access business, positioning themselves as service providers and complementing their traditional sales and network operations with a third element, a "telco innovation factory" charged with developing and marketing new services. The innovation factory consist of access-centric services that use the existing network and IT platforms – in the e-health segment, for instance – or regional OTT-related offerings such as TV. Such services will be embedded in partners' service suites or, depending on the extent to which the ad valorem aspect is to be emphasized, on proprietary platforms that integrate third-party services. SingTel is attempting to buck the trend of telcos becoming just a big, fat dumb pipe that only competes on price. Their vision is to really go after the heart and soul of the consumer, ultimately to drive a deeper connection that substantially increases the value of SingTel.

Aligning strategic positioning with shared values opens up fundamentally new ways of thinking about business. From being known as a telecom carrier, Singtel can become a

smarter cities builder, a wellness provider or a health and safety enabler : whatever !!
They have recognized that mobile operators will play a crucial role in working together
with a range of industry partners in health, automotive, education, smart cities and a
range of vertical industries to accelerate the launch of valuable connected services.
SingTel Life Labs is a global innovation initiative to accelerate innovation and application
development through collaboration with strategic partners, renowned research institutes
as well as the innovator and developer ecosystems. They created an App that helps
Singapore residents navigate increasingly large and confusing shopping malls using "
sensor fusion " and Wifi triangulation .

Chua Sock Koong, SingTel Group CEO, states : "SingTel has a long history of quietly, but
successfully, making bold and industry-shaping investments. We now see some of the
largest and most exciting opportunities that have ever existed in this industry... We need
to learn from past failures and be prepared to reiterate a bold idea if we believe it will
eventually bear fruit. Even when we do not succeed, we expect a "fail fast and fail cheap"
mentality to produce valuable learnings that can form the basis of long-term advantage
against competition.."Strategic partnerships , savvy aquisitions , superior digital insights ,
superfast networks : this kind of cerebral leadership summarises why SingTel has
become such great company. And by the way ...their CEO...Chua Sock Koong is a Lady
who has shown that women can succeed in traditionally a male dominated telco world !!!

--♠--

Telco Digital # 2 : The Technology Agenda

"Communication networks are facing a lack of scalable and sustainable architecture to
meet the challenges ahead in terms of data traffic increases, video uploads and
downloads, and enhanced M2M communication. The network of the future has to be
highly elastic in order to facilitate the adding or dropping of capacity and real-time
provisioning of service. It needs to be highly orchestrated by key business imperatives,
such as customer satisfaction, and it must be highly integrated so that synergies are fully
embedded and captured across fixed and mobile, across borders and across segments."

(Bruno Jacobfeuerborn Deutsche Telekom CTO)

Software Defined Networking or SDN is a technological approach to designing and managing networks that has the potential to increase operator agility, lower costs, and disrupt the vendor landscape.With SDN the network becomes a programmable fabric that can be manipulated in real time to meet the needs of the applications and systems that sit on top of it. SDN promises fully automated, application-aware and adaptive adjustments to bandwidth, compute power and storage with end-user visibility is what is needed to provide ultimate QoE to the mobile user connected an IP based Telecom network such as LTE .

The root cause of a network's limitation is that it is built using switches, routers and other devices that have become overly complex because they implement an ever-increasing number of distributed protocols and use closed and proprietary interfaces. By decoupling the network control and data planes, OpenFlow-based SDN architecture abstracts the underlying infrastructure from the applications that use it, allowing the network to become as programmable and manageable at scale as the computer infrastructure that it increasingly resembles. An SDN approach fosters network virtualization, enabling IT staff to manage their servers, applications, storage, and networks with a common approach and tool set.

In a SDN, the network administrator can shape traffic from a centralized control console without having to touch individual switches. The administrator can change any network switch's rules when necessary — prioritizing, de-prioritizing or even blocking specific types of packets with a very granular level of control. This is especially helpful in a Cloud computing multi-tenant architecture because it allows the administrator to manage traffic loads in a flexible and more efficient manner. SDN allows network engineers to support a switching fabric across multi-vendor hardware and application- specific integrated circuits.

Most of the churn in mobile subscribers today is attributed to poor QoE (Quality of Experience). What's needed is an efficient and elastic system that adapts to the end-user traffic automatically and dynamically. The rapid adoption of 4G Mobile (LTE) necessitates uninterrupted availability of quality services 24/7 regardless of location or device. SDN has the capability to make this a reality. The Open Flow protocol allows the network to be programmed on a per-flow basis and thereby provides visibility at the user and application level. The capability to increase or decrease the bandwidth needed, for instance, by way of automated bandwidth signalling is one advantage. It can also adjust the number of VMs (Video Messages) and the associated storage needed proactively and dynamically with¬out any human intervention on an application basis.

Developing an SDN business involves the deployment of physical infrastructure, a network controller and a telecoms operating management system which combines operation and business support systems. The network controller is central to SDN with two main functions: virtual resource control and traffic management systems (TMS). The network controller can create a programmable, logical network that allocates resources within the physical network (access and core networks) in the most dynamic way without needing to know the actual infrastructure topology. In so doing, the operator can build the most appropriate virtual network offering multiple services.

SDN is not only an esoteric technology concept but a current reality: in 2012 Google announced that it had migrated its live data centres to a Software Defined Network using switches it designed and developed using off-the-shelf silicon and OpenFlow for the control path to a Google-designed Controller. Google claims many benefits including better utilisation of its compute power after implementing this system. Recently Japanese vendor NEC established a partnership with Portugal Telecom that will see the two firms collaborate on SDN (software defined networking) and virtualisation technology for data centers and carrier networks. The two firms claimed the agreement would enable both companies to test and assess the commercial feasibility and benefits of SDN implementation for carrier data centers , adding that SDN and network virtualisation have "exceptional potential".

According to Infonetics, telecoms plan to deploy SDNs and NFV by 2014 within data centers, between data centers, operations and management, content delivery networks (CDNs), and cloud services. In most cases, Telcos are starting small with their SDN and NFV deployments, focusing on parts of their network, in " contained domains " such as data centers, to ensure they can get the technology to work as intended.Running in parallel Telco network architects believe that NFV (this term Network Function Virtualisation was coined by the European Telecommunications Standards Institute) will consolidate many network equipment types onto industry standard high volume servers, switches and storage, thus providing a new network production environment so as to lower cost, raises efficiency and increases agility. Network Functions Virtualisation can be implemented without the prescence of a SDN , although the two concepts and solutions can be combined to unlock greater value.

In the next decade SDN is big business. According to SDN Central (an independent market research community for SDN & NFV), the SDN market is expected to surpass $35 billion in the next 5 years. Adoption of SDN technology has accelerated in recent years from sales of $10 million in 2007 to $252 million in 2012.The emergence of the software-defined networking market is supported by growth in venture capital investment in SDN focused companies. Venture capital funding rose from $10 million in 2007 to $454 million in 2012.

The availability of high bandwidth wireless packet technologies coupled with competitive flat rate pricing has fuelled an explosion in mobile broadband. Analysts predict that within 4 years 75% of mobile data traffic will be driven by video, but are Telco networks ready to cope? Are Telcos positioned to make money from this explosion in demand or will it be driving even revenues to the OTT players ? Peer-to-peer video download applications, for example, can quickly consume network resources and impact the performance of the overall network.

In the era of rapidly-increasing bandwidth demand but flat to decreasing ARPUs, service providers need smarter wireless networks to improve margins. These smarter networks can offer service providers greater control of their network to optimize operational costs and enable new revenue streams. Users are demanding a similar media-rich broadband internet experience in the wireless world as they have in the wired world. The impact of mass market mobile broadband has a severe impact on the underlying wireless core network.It neccessitates the deployment of next generation of intelligent packet core technology.

Referred to as the intelligent packet architecture mainly due to the flexible and flow based policy and charging mechanisms that are integrated into the GGSN these intelligent mechanisms leverage Deep Packet Inspection technology to enable services such as Content Based Billing and precise QoS and bearer resource management. Consolidating everything into a single intelligent packet core node provides for capex savings as well as a more coherent single point of awareness approach to delivering IP services. This new generation architecture facilitates reliable, scalable, high density packet processing capability to meet the challenge of the data / video tsunami while reducing the overall cost of ownership for the operator The increased end user bandwidth available with mobile broadband technology is the resource for the sale of new revenue generating incremental services. These new services can be coupled with flexible flow based charging and DPI based content awareness to provide new and innovative marketing-led service offerings and charging schemes.

Understanding per-subscriber content traversing the wireless network is the critical first step to identifying problem areas as well as potential revenue opportunities. Which subscribers are consuming the greatest amount of bandwidth? Which applications have the biggest impact on your network? What are the time-of-day usage patterns? Fully understanding the usage of the network at a granular level is an invaluable tool to helping identify problem spots, causes of congestion, and which subscriber applications offer the greatest potential for profitability. Content management provides the answer to this business model challenge.

Content management platforms provide the means for operators to monetize the open Internet, make new revenues and, at the same time, optimize the cost of running these networks, making the service more profitable. It helps a network to transition from being a dumb bit-pipe to an intelligent profit-making network. In addition the use of an Intelligent S-GW (LTE) with its ability to manage bandwidth and deep understanding of real time content enables service providers to create unique and differentiated services.

Leveraging insightful reports and charts generated by the gateway management software, wireless operators will gain a deeper understanding of how their network is being used, where congestion problems may arise, and what revenue opportunities. This valuable information can be used for marketing, product planning or network control. Operators are able to extract a very granular view of content traversing the network on a per-subscriber, per-flow basis including hard-to-detect traffic, such as peer-to-peer file sharing or Skype, as well as standard applications such as email and video streaming.

The Gateway function supports extensive accounting and billing options for service providers, enabling them to effectively capitalize on service differentiation opportunities. In addition to traditional time and volume based charging the Gateway enables more flexible and intelligent charging solutions such as Content Based Billing (CBB), Flow Based Charging (FBC) as well as tiered and event based charging. Content Based Billing leverages the Deep Packet Inspection technology within the AGW / SGW while Flow Based Charging builds on the service data flow granularity provided by a 3GPP PCC compliant architecture.

The open Internet and rich mobile broadband model presents unique challenges to operators as they strive to effectively manage network resources and drive additional revenue and profit. Unfortunately , flat monthly rates with lowering ARPUs and the higher cost of burgeoning bandwidth consumption makes the mobile broadband business model challenging. Content management platforms help a network to transition from being a dumb bit-pipe to an intelligent profit-making network.

Rather than offering flat rate billing for opaque dumb bit pipes, operators can offer "smart" pipes that provide premium QoS for specific applications, such as video streaming or music downloads. The evolved packet core gateways leverage feature-rich, services at the edge architecture together with a full set of standardized and open interfaces. This delivers the key policy, charging and security enablers needed for service differentiation whilst protecting subscribers and networks from the threats posed by the new world of mass market mobile broadband.

As all Telco engineers know that in a typical mobile deployment, each base station serves all the mobile devices within its reach. Each base station has its digital component manage its radio resources, handoff, data encryption and decryption and an RF component which transforms the digital information into analog RF. The RF elements are connected to a passive antenna that transmits the signals to the air. Each base station should be placed in the geographical center of its coverage area. But even when such locations are selected, the mobile operators may have difficulty in renting the real estate, finding proper powering options, securing the location and protecting the equipment from weather conditions. Those cell sites carry with them a continuous stream of OPEX to address the high rental rates for real estate, electrical expenses, cost of backhaul for the cell site and security measures to protect the location from intruders.

Enter a novel architectural paradigm : C RAN !!! The basic premise of Cloud RAN is to change the traditional RAN architecture so that it can take advantage of technologies like cloud computing, Software-Defined Network (SDN) approaches, and advanced remote antenna/radio head techniques.C-RAN architecture is not bound to a single RAN air interface technology. In essence, conventional terrestrial cell site base stations are replaced with remote clusters of centralized virtual base stations which can support up to a hundred remote radio / antenna units. This is achieved by centralizing RAN functionality into a shared resource pool or "cloud" (the digital unit – DU, or baseband unit – BBU) which is then connected via fibre to advanced remote radio heads ("Radio Units" – RU) sited in different geographical locations in order to provide full coverage of an area. The radical concept can even use banks of x86 servers to connect cellular calls rather than traditional wireless base stations.

From a business perspective,C-RAN will deliver significant reductions in Opex and Capex due to reduced upgrading costs. A major reason for this is the aggregation and pooling of the DU computing power which can be assigned specifically where needed e.g. the load situation over time and space for indoor/outdoor cells, am/pm hours, weekday/weekend, and so on. As a result, single cells do not need to be dimensioned for peak hour demands, but rather the processing power can be pooled and assigned on an on-demand basis. The processing power savings achieved should also leave processing headroom for any further potential technology enhancements (e.g., LTE-A features) without the need for further CAPEX. C-RAN skips the need for a high-bandwidth, low latency (X2), synchronized interface between the geographically distributed base station because the computing resources of the multiple transmission points' BBUs are all located within the same hardware.C-RAN slashes capex because fewer BBUs are needed, which reduces opex because fewer BBUs means less energy consumption and diminished maintenance costs. The reduced energy consumption makes C-RAN a "green" alternative, with China Mobile estimating 71 percent power savings vs. traditional RANs.

Furthermore, interference management will also benefit from C-RAN network architecture as technologies like dynamic eICIC schemes will be enabled, especially in a HetNet deployment .Heterogeneous networks will require small cells to be independent, intelligent and ubiquitous to avoid the cross- interference mayhem, yet be in synch and orchestrated with macro cells (including Cloud – RAN topology).Small cells are poised to become the most commonly used node for cellular access in the next-generation HetNet. C RANs will likely take their place beside traditional base stations and emerging small-cell base stations as another tool for building cellular nets.The success of many new 4G network deployments will depend on the use of outdoor and indoor small cells to extend coverage and increase capacity in areas poorly served by macrocell networks. Operators are also considering proposals to deploy more efficient CloudRAN architectures requiring high speed CIPRI front haul links between remote radio heads and pools of baseband units.

According to Maravedis Cloud-RAN economics only be realized by harnessing standards to ensure interoperability and reduce cost. That, in turn, will create a whole new ecosystem, and operators must resist any attempts by their suppliers to hijack standards for software-defined networking or cell site equipment. Otherwise, this fledgling architecture will remain confined to a few pioneers with the resources to build their own ecosystems, like China Mobile. China Mobile, the world's largest carrier with 700 million subscribers, has been spearheading trials and plans to deploy systems as early as 2015. Japan's NTT Docomo said it will follow in 2016, and a third unnamed carrier is now preparing plans for C-RANs. China Mobile aims to lower the cost of C-RANs to less than $30 per LTE sector, down from about $10,000 two years ago. It will start a second round of trials later this year using servers equipped with PCI Express cards to handle baseband processing. Each card will pack four FPGAs using silicon cores, each FPGA capable of handling 12 LTE sectors.

As MNOs face rising CAPEX bills to meet mobile data demand combined with falling ARPU, they must explore radical new network designs. With Cloud-RAN, they can virtualize baseband processing functions for hundreds of sites on a server or base station hotel. By consolidating individual Base-station processing into a single or regional server farm Investments on Cloud Radio Access Network (RAN) Infrastructure are expected to exceed $6 Billion by 2020, according to a new report from SNS Research. Distributed antenna technologies (DAS) will get a new lease on life, supporting coverage extension for C-RAN sites. This sector will open up $1.3bn in new revenues for antenna providers.

Pure C-RAN faces many barriers, such as over-reliance on fiber to link sites and basebands and immature standards, but most operators will inch towards C-RAN using hybrid models. Development of microwave fronthaul technologies will be critical to improve the C-RAN business model . Whatever the challenges C-RAN offers a revolutionary approach to next-generation cellular networks deployment, management and performance. Fiber, needed for fronthaul, is crucial to C-RAN deployment, so it is no wonder that fronthaul is constantly brought up as Cloud RAN's biggest challenge. Fronthaul connects RRHs to the aggregated BBUs, with traffic then backhauled from the BBUs to the IP core or evolved packet core (EPC).

NTT DOCOMO, Japan's leading mobile operator and provider of integrated services centered on mobility, announced today it will begin developing high-capacity base stations built with advanced C-RAN architecture for DOCOMO's coming next-generation LTE-Advanced (LTE-A) mobile system. The new architecture will enable quick, efficient deployment of base stations, especially in high-traffic areas such as train stations and large commercial facilities, for significantly improved data capacity and throughput.Advanced C-RAN architecture, a brand new concept proposed by DOCOMO, will enable small "add-on" cells for localized coverage to cooperate with macro cells that provide wider area coverage. This will be achieved with carrier aggregation technology, one of the main LTE-Advanced technologies standardized by the Third Generation Partnership Project (3GPP). The small add-on cells will significantly increase throughput and system capacity while maintaining mobility performance provided by the macro cell.

For NTT DoCoMo high-capacity base stations utilizing advanced C-RAN architecture will serve as master base stations both for multiple macro cells covering broad areas and for add-on cells in smaller, high-traffic areas. The base stations will accommodate up to 48 macro and add-on cells at launch and even more later. Carrier aggregation will be supported for cells served by the same base station, enabling the flexible deployment of add-on cells. In addition, maximum downlink throughput will be extendible to 3Gbps, as specified by 3GPP standards.C-RAN is typically thought of as a large-scale urban macro solution, but the concept of pooled baseband serving n number of radio access nodes can apply to a variety of scenarios, such as small cell underlays (using micro RRUs), so-called Super Cells, and outdoor/indoor hotzone systems. These models, identified and defined partly through the NGNM Alliance, could prove an attractive way to introduce and develop C-RAN technology. Given the traditional RAN's coverage restrictions and limitations of transmission and reception signal support, the benefits of deploying a C-RAN infrastructure are clear.

Bottom Line : The C-RAN, as a centralized, general purpose processing solution, enables the efficient use of network resources. Based on open-platform and base station virtualization, C-RAN provides an ideal architecture for LTE-A functionality as well as being complementary to next-generation SDN and NFV deployments. Many major mobile operators across the globe are preparing to incorporate the cloud into their existing RAN platforms. We anticipate that 2014 will move the C-RAN beyond the "cloud hype" as operators gain a competitive edge through integrating the C-RAN in their LTE-A migration.

By 2017, it is expected that LTE will account for about one in eight of the more than eight billion total mobile connections forecast by that point, up from 176 million LTE connections at the end of 2013.GSMA research also reveals that nearly 500 LTE networks are forecast to be in service across 128 countries, roughly double the number of live LTE networks today. Despite the explosive growth in LTE this year i expect 5G to take centre stage....even though the first commercial networks will be rolled in 2020....eventhough the standardisation process is half done even though many of the building blocks of 5G-technology architecture are as yet unknown. Now you may wonder why is 5G causing us cerebral thrills ?

If you were blissfully unaware ...we are at the dawn of an era in networking that has the potential to define a new phase of human existence. This era will be shaped by the digitization and connection of everything and everyone with the goal of automating much of life, effectively creating time by maximizing the efficiency of everything we do, augmenting our intelligence with knowledge that expedites and optimizes decision-making and everyday routines and processes. According to the venerable Bell Labs there are 3 catalytic technological drivers for this future 5G Network:

 1:A new dynamic approach to networks that creates the sense of seemingly infinite capacity by pushing beyond current scientific, informatics and engineering limits to create a new cloud integrated network (CIN) that not only provides the essential input and output mechanisms, but also, intelligence.

2:The rise of Internet-connected machines and devices that will send and receive a massive amount of new digital information, as well as reshape the manufacturing landscape with a diverse array of 3D printers — reversing the current trend in developed countries of off-shoring manufacturing to lower-cost countries.

3 :New data analysis techniques based on inference of needs and information, instead of seeking "perfect knowledge". New augmented intelligence systems will use the smallest amount of digital information, derived from the massive amount of data collected, to infer what is needed in each situation and context, to assist – not replace – human intelligence.

So we are entering an era where potentially 1000x more capacity is required, and as such we are obliged to take a different approach to networking, by exploring new architectural dimensions.To achieve 5G by 2020 is heavily dependent on Moore's Law when it comes to new computing platforms. Increase in device performance will continue at rates predicted by Moore's Law because the number of transistors and IC's doubles about every two years.

So expect new gear that enables embedding the cloud in the network to form a new edge cloud to provide the optimum performance (throughput and latency) and economics for both virtualized networking functions and any other performance critical enterprise or web services. Anticipate the emergence of an end-to-end, software-defined networking layer that dynamically connects distributed and diverse workloads, networks and devices, and creates end-to-end virtual network paths or slices . And finally behold the creation of a new ultra-high capacity and continuously reconfigurable network fabric, with the reimagining of the core, metro and access layers and architecture.

Did you catch my drift amigo ? Its software that will gobble up the network of the future. Non-quantitative capabilities of 5G technology include a soft ware- based system architecture, simplified authentication, support for shared infrastructure, multi -tenancy and multi -RAT (with seamless handover), support for terrestrial and/or satellite communication, robust security, privacy, and lawful interception capacity.In a nutshell 5G will provide an order of magnitude improvement in performance in the areas of more capacity, lower latency, more mobility, more accuracy of terminal location, increased reliability and availability. 5G will allow the connection of many more devices simultaneously and to improve the terminal battery capacity life. The enhanced spectral efficiency will enable 5G systems to consume a fraction of the energy that a 4G mobile networks consumes today for delivering the same amount of transmitted data. 5G will reduce service creation time and facilitate the integration of various players delivering parts of a service.

Consider the healthcare industry in which hospitals can arrange remote robotic surgeries via a customized 5G network that minimizes network latency as if the surgeon were physically present next to the patient. Or how skin-embedded and 5G connected healthcare chips could constantly monitor vital signs, prevent conditions from becoming acute, and constantly adapting medication to meet changing conditions.

With sensors enabled by 5G networks, every water pipe could be monitored in real-time and utility providers could create a network that can sense, process and transmit exact locations and severity of a leak and alert proper resources in real time without the need for humans to laboriously collect and analyze the data. Similar 5G-enabled transformations are only to be expected in agriculture, finance, retail, education, trade and tourism. The possibilities are truly endless.

Last but not least, 5G will open the ecosystem for technical and business model innovation. The extension of the cloud computing model to the telecom industry will unleash innovation and allow new players to access the ecosystem. Since 5G, network services will rely overwhelmingly on software it will strengthen local software industry, including SME developers and solutions providers that can better compete in an increasingly hardware-agnostic market.

We all know that Radio based services rely on appropriate access to EM spectrum at suitable frequencies. To meet the expected growth in traffic and requirements associated with radical applications such as Virtual and Augmented Reality , IoT , Autonomous vehicles... the success of 5G systems and services depends entirely on a more efficient use of spectrum already assigned to terrestrial mobile services; and the timely ability to utilise certain new bands in order to support new capabilities for which demand exists.Spectrum flexibility can bring benefits of spectral efficiency gains, examples include: increasing exclusive spectrum with emphasis on improving regional/global harmonisation; smart carrier aggregation to use spare frequencies; spectrum trading; and managing fair access to supplementary shared spectrum.So what can we expect on the Spectrum front for 5G ?

The Regulators in Europe , USA , Korea , Japan and China are already investigating the feasibility of wide contiguous bandwidths, higher carrier frequencies above 6 GHz even as networking vendors accelerate their research on new wave forms to push Shannon's Law to the limit. Whilst access to additional spectrum above 6 GHz is of interest, it should be emphasized that in general low frequency spectrum (below 6GHz), especially sub-1GHz, is still absolutely essential for an economical delivery of mobile services and this holds true for existing systems as well as future 5G systems.The many socio economic benefits of 5 G are entirely dependent on the Spectrum Regulators and Governments.

Supplementary spectrum, made available on a shared basis, will be required to deliver extra capacity where needed, for example additional licensed spectrum made available by an incumbent governmental / public user within a defined geographic area and/or defined time. Access to licence-exempt spectrum as a useful supplement for certain applications and will be seamlessly integrated into the 5G platform.

Pundits say 5G will change the world...you bet. But these positive changes will be felt unevenly across the planet. Infact it might make the Digital Divide even worse. The First world (USA , Europe , Korea , Japan et al) will rollout their 5 G because they will have spectrum to do it to enjoy light speeds . And the so called 3rd world ?...they are still battling to get spectrum for 4G so what of 5 G ? So I expect proactive Regulators will make it happen in the First world but break it in the Third world unless the latter get their act together NOW.

Believe it or not 5G is hurtling at us like the Starship Enterprise at warp 7. Tier 1 Telcos such as Verizon , NTT DoCoMo, BT , Etisalat etc are busy trialling with the main vendors in Test Beds and research universities are achieving speeds of I Tbs. Thats your HD movie download in 4 seconds. Governments from EU,Korea , Japan are throwing millions into research even as respective Regulators get ready to dole out Spectrum to utilize higher frequency spectrums ranging from 6 to 66 gigahertz. This would take it into the millimeter-wave band, which will enable multi-beam multiplexing and massive multi-input-output (MIMO) technologies.In future, 5G will have a much wider influence in areas such as realtime monitoring of IoT sensors, due to its low latency attributes. with the arrival of the 5G era marked by Gbps-level speeds and ultra-low latency, we will witness a dramatic improvement in the content and service quality along with increased number of content and service channel..... If you are a Star Trek fan, then 5G-powered Internet of Things (IoT) and Virtual Reality (VR) should fulfill some of your dreams..Beam me up Scotty !

M2M market (connecting devices other than phones, laptops and similar consumer devices is the new gold mine for network operators seeking data revenues. According to the international research firm Gartner, M2M is one of the Top10 mobile technologies to watch. Berg Insight predicts the number of cellular connections used for machine-to-machine communication to grow at a compound annual growth rate (CAGR) of 25.6 percent to reach 187.1 million worldwide by 2014. In the same timeframe, M2M's share of the global cellular network will rise from today's 1.4% to reach 3.1%

Telcos must understand that while mobilizing the machine is based on the same cellular network as that used for voice services, in most other respects the offering differs. Characteristics such as time of usage, frequency of usage, file size, and customer base will all be different. Consequently, rate structures, sales channels, service levels, and technical support must all be re-designed with embedded mobile in mind. M2M (sometimes called Embedded mobile) is likely to generate a lot of additional data traffic, so it will be an important potential source of revenue and growth. Particularly if some of the data traffic can be steered toward the network's off-peak periods. Network infrastructure is a sunk cost, and operators have predictable periods when they know the network will be under-utilized. According to Machina the market that is addressable by CSPs is $339 billion but of that only $5 billion will be about connectivity. But connecting it all, and connecting it at a level of service that CSPs need to guarantee must happen as stage one. As with people, CSPs need to choose their opportunities. Low ARPU connectivity may be a model for one CSP, value added services might be the model for another.

Amongst the pressing issues are standards adoption, cloud based service delivery platforms and security. With M2M network operators will probably not own the customer relationship as they do with voice. M2M applications are too complex and specialized for most companies to develop and deliver on their own. Instead, different players will need to collaborate to provide a total solution. For example, creating a building-automation system involved a range of different actors, including the chip and RF-chain vendors, the device assemblers, the operating system vendors, the network operators, and the application developers.Sample M2M applications include but not limited to :

Smart Metering : Smart Metering has become an essential part in the public utility industry. With government regulations and cost increase of nature resources (crude oil, coat, nature gas,etc), public utility companies have to implement an automatic real-time meter reading process to save cost and time. The real-time utilities usage information not only enables providers to allocate resource more efficiently but also allows customers to obtain more accurate billings.

Sales and Payment :Electronic payment system becomes the essential portion of people daily livings. More and More point-of-sale (POS) terminals, nowadays, are embedded cellular wireless modules in order to operate independently anytime and anywhere. Cellular wireless POS terminals can also feedback real-time sale information to vendors for resource allocation and sale & marketing analysis.

Security : The concern of safety is raising sharply. The demand of alarm systems to be installed in private residential, commercial and public locations grows exponentially. Cellular technology is applied to organize the usage of alarm systems according to different desires. Law-enforcement units, emergency organizations and private individuals can react more effectively upon the real-time alerts of various occurrences.

Healthcare: With cellular technology enable device, patients no longer have to be present in physician clinics to obtain medical advice. Physicians now are able to monitor their patients' condition continuously regardless of the local of the patients. Unnecessary clinical visits can be eliminated. Healthcare system will become more effective and efficient.

Telemetry : Shipments of medicine or food require the observance of certain temperature ranges or a complete documentation of the cold chain. Sensors inside the containers measure parameters such as temperature and humidity and transfer them to a central database using M2M technology.

Telcos are sitting on a gold mine of core capabilities and assets, critical for M2M success. In addition to just network connectivity, they have outstanding expertise in large scale service delivery with high reliability and global reach, plus an arsenal of partners for content, apps, specialized solutions and services, integration, as well as terminal and module vendors.While the price of devices has been steadily decreasing, mobile operators all over the world have been driving growth in the M2M market over the past several years as we can see by rapidly increasing numbers of M2M subscribers and higher revenues.

Telia Telenor is the Telco industry's poster boy in providing M2M applications based on their of horizontal, multi-tenant platform for M2M .The design of their Telenor Objects' software platform is a layering concept which provides a middleware layer to help different kinds of devices, networks, and applications to interoperate. This facilitates open interfaces for data capture, data pre-processing, device management, and information exchange with other systems. Telenor Objects is based on five key elements : the brand and channels to market, the technology, the partners, a managed-service business model drawing on the telco heritage, and open-source software.The complete system provides a range of reuseable capabilities, including a GUI for users, developer APIs, and a device library. The system core is a software platform which implements a secure message-exchange and device management system.

Sprint launched its Emerging Solutions Group in 2009. Some of the M2M offers include remote monitoring to keep track of Alzheimer's patients and providing news and weather updates for digital signage systems. They are also involved in remote monitoring and control of equipment, primarily SCADA units for oil wells and waste water, as well as security, utility meters and appliances. Some of their partners include Grid Net, Landis+Gyr, and Ford Motor Company.Verizon is targeting enterprise clients with wireless ATMs, electronic medical records, RFID-tagged supply chain applications, video surveillance, stolen vehicle tracking, monitoring of water, gas and electric distribution, control electronics, DVD kiosks and vending machines.

Orange is focusing on vehicle tracking and online monitoring and reporting, asset management, wireless CCTV safety cameras, satellite navigation systems, patient medical trials, patients' diaries, security, asset tracking, dynamic signage, vending machines, stock control, pest control, remote measurement of water and energy usage, automatic meter reading, wireless alarm systems, emergency lighting and wireless closed circuit television apps.

However, the road ahead is not easy. Continuous success for operators in this space will require transformational strategies along with new value propositions, product innovation and perhaps even some mergers and acquisitions. At the same time, operators need to have good insight into the software and system solutions that are emerging as the fastest growing segment in the M2M industry. Increasingly advanced large-scale M2M applications require advanced service enablement platforms that integrate remote devices, mobile networks and enterprise applications.

The challenges involved in exploiting M2M are significant but the potential benefits are even greater. For enterprises that use M2M embedded mobile, there is the potential for greater efficiency, improved business processes, and innovative new business models. For network operators, mobilizing the machine represents a way to greatly extend the subscriber base and drive up data traffic. For cellular device vendors, machines are a vast new target market. For OEMs, embedding mobile in their products allows differentiation, has the potential to expand their customer base, and could even enable new product lines.

Carrier aggregation is one of the most distinct features of 4G systems including LTE-Advanced, which is being standardized in 3GPP as part of LTE Release 10. This feature allows scalable expansion of effective bandwidth delivered to a user terminal through concurrent utilization of radio resources across multiple carriers. These carriers may be of different bandwidths, and may be in the same or different bands to provide maximum flexibility in utilizing the scarce radio spectrum available to operators.

The key to achieving higher data rates with LTE is to enable network operators to use the technology in bandwidths wider than 20MHz. Some network operators may be lucky enough to have contiguous spectrum allocations of more than 20MHz. However, the nature of spectrum allocation over the years is such that most operators have a mix and match of spectrum within and between frequency bands. Following the redistribution of analogue TV spectrum and the provision of higher frequency spectrum, an operator might have LTE spectrum at one or more of 700MHz, 800MHz, 900MHz, 1800MHz, 1900MHz, 2100MHz, 2500MHz and 2600MHz.

Recently Telekom Austria, conducted a successful demonstration of LTE-Advanced carrier aggregation .The live demo, which included handling large file transfers with a simultaneous video stream, showcased download speeds of 580 megabits per second (Mbps), far more than twice the current 4G LTE peak rates. In combination with carrier aggregation, Telekom Austria plan to turn LTE-Advanced into a Gigabit technology that will allow a large number of users to simultaneously access high data rates within one mobile radio cell.

The U.S. market is one of the main drivers for deployment of Carrier Aggregation since the frequency spectrum is a scarce resource and very fragmented with few operators having contiguous 20 MHz spectrum generally available. The LTE Carrier Aggregation roll out will start during 2013 and we expect significant growth 2014. U.S., Korean and Japanese operators already have firm deployment plans for Carrier Aggregation. In the first phase, up to 20MHz of spectrum will be aggregated enabling subscribers to enjoy up to 150Mbps downlink data throughput, or even higher in the future.CA (Carrier Aggregation) may be used in three different spectrum scenarios :

Intraband Contiguous CA — This is where a contiguous bandwidth wider than 20 MHz is used for CA. Although this may be a less likely scenario given frequency allocations today, it can be common when new spectrum bands like 3.5 GHz are allocated in the future in various parts of the world. The spacing between center frequencies of contiguously aggregated CCs (Component Carriers) is a multiple of 300 kHz to be compatible with the 100 kHz frequency raster of Release 8/9 and preserving orthogonality of the subcarriers with 15 kHz spacing.

Intraband Non-Contiguous CA — This is where multiple CCs belonging to the same band are used in a non-contiguous manner. This scenario can be expected in countries where spectrum allocation is non-contiguous within a single band, when the middle carriers are loaded with other users, or when network sharing is considered.

Interband Non-Contiguous CA — This is where multiple CCs belonging to different bands (e.g., 2 GHz and 800 MHz are aggregated). With this type of aggregation, mobility robustness can potentially be improved by exploiting different radio propagation characteristics of different bands. This form of CA may also require additional complexity in the radio frequency (RF) front-end of UE. In LTE Release 10, for the UL the focus is on intraband CA, due to difficulties in defining RF requirements for simultaneous transmission on multiple CCs with large frequency separation, considering realistic device linearity. For the DL, however, both intra and interband cases are considered in Release 10, while specific RF requirements are being developed.

Whereas GSM made use of only four frequency bands globally and WCDMA/HSPA requires five bands to get global coverage, LTE is deployed on more than ten bands already as of December 2012, and the number of bands will still increase. The frequency band fragmentation is a result of regional decisions on how the spectrum shall be utilized. Of course, there are initiatives to define some bands as global roaming bands. Another big challenge for Carrier Aggregation is to develop a RF front-end flexible enough to support a majority of domestic, and international, bands together with possible future Carrier Aggregation combinations

LTE Carrier Aggregation will provide mobile network operators with even greater scope to support services that hitherto would have been restricted to fixed networks, and may open up the possibility of providing a viable alternative to fixed network broadband services, particularly in rural locations where fixed broadband provision may be poor. Which is why CA is extremely important for Telcos in developing countries that are battling their sluggish Regulators for LTE Spectrum ?

At the current rate of LTE Spectrum allocations it is most unlikely that the long suffering Third World broadband consumer will ever experience the high speed LTE. By the time Regulators release the Digital Dividend spectrum in Africa (2015 – 2018), LTE Advanced will be a globally standardised and mature technology. This is the silver lining, believe it or not , since Telcos in developing countries will have the opportunity to launch LTE Advanced with all its speed benefits and refinements like Carrier Aggregation by the time they get some Spectrum.

Enterprises struggle to store and manage Big Data because it exceeds the capacity of current relational systems and the reason is clear: those legacy systems were designed decades ago, long before Big Data was front and center in the collective imagination. For Telcos, the velocity of data growth and increasing subscribers mean traditional data analytics software will take months to process information which is needed in real-time. Existing tools included operational-type reporting, looking at log files and extracting information from them. The more critical mobile data services have become to the business, the greater the need has become to monitor their contribution. This means being able to track numbers of 'unique active users' of its various services, information that had not been easy to come by previously. Today's relational databases and business intelligence tools are powerful, but what if you have 100, or 1000 times the data?

Enter Hadoop !! No not the toy elephant but a cloud-based, open source platform capable of mining Big Data ona vast scale by harnessing huge arrays of inexpensive computer processing power. Hadoop is an open-source software framework for storage and large scale processing of data-sets on clusters of commodity hardware. It is engineered to spread out that processing across hundreds if not thousands of plain vanilla servers (and eventually, in Google's vision, millions of machines) arranged in a cluster, rather than relying on super-expensive proprietary machines.

Hadoop is an Apache top-level project being built and used by a global community of contributors and users. Hadoop makes it possible to run applications on systems with thousands of nodes involving thousands of terabytes. Its distributed file system facilitates rapid data transfer rates among nodes and allows the system to continue operating uninterrupted in case of a node failure. This approach lowers the risk of catastrophic system failure, even if a significant number of nodes become inoperative.

By turning storage and processing into a commodity, Hadoop allows organizations to be more nimble and agile. SQL-based Hadoop could be used by SQL users, or a new product team could instead use scalding or standard Unix Streaming to achieve results, or a data scientist could use Python libraries that they have come to depend on, all in the same framework and more importantly on the same infrastructure.And what does this all mean for the traditional telecom service provider, not to mention those companies' hosting/cloud computing groups? Plenty !!! Because not only is Hadoop processing likely to be an important application for cloud providers, telecom operators — sitting on reams of network and customer data — are prime candidates to become Hadoop users.

China Mobile are using Hadoop as a telecom data mining platform, showing how operators can tap this powerful technology to better understand their networks, services and customers, finding new patterns and revelations that can help them compete in the digital future. China is more open to open source than proprietary software because of capabilities that can be gained from an engineering perspective.

The government of the Chinese province of Zhejiang is tapping Hadoop to generate about 2.5 petabytes of data each month in the form of video streams from CCTV cameras. With Hadoop, the Zhejiang government was able to solve the big data problem of storing, monitoring, searching ,and analyzing the data in real-time.In India, a mobile advertising firm used Hadoop to help its telco customers deliver relevant real-time information to subscribers. For example, a subscriber who has tried buying a product when they do not have enough prepaid mobile credit will be reminded to make the purchase after they reload credits into their accounts.

Google uses Hadoop (actually its own proprietary version, called MapReduce) to help it swallow the entire Web, not to mention massive map/satellite databases, to produce elegantly useful products such as Google Maps, Google Earth and, of course, the Google search engine itself. Yahoo! uses Hadoop to analyze and optimize how its 20 million visitors consume its home page content.The New York Times set a Hadoop-powered cloud against an 11-million story archive dating back to 1851 to make it instantly searchable.Facebook uses Hadoop to analyze interactions and social graph links on its site — growing at a rate of more than 20 terabytes of new data per day — powering the friend connections and personalization that drives the social networking site.

One of the largest mobile network operators in Europe wanted to leverage the data it captured on mobile usage to achieve a number of specific benefits. Managing advanced data services such as packages that provide mobile Internet access on a range of devices, and applications including mobile email, instant messaging, Google search, news and sports updates, and weather and traffic reports was a monumental and expensive task.Accurate data about real customer activity would help drive changes to its portal, giving users easy access to the applications they use most often. Analysis of service usage enables the company to spot upcoming trends and intelligently market them to customers, and keep its customer touch points current.

After Hadoop implementation the individual business units can report on usage, and provide other KPI information because it allows huge amounts of data to be stored in a granular fashion that is cost effective and performance. With greater and more granular visibility, the company is able to keep its mobile applications and web portals fresh and in line with current customer interests, increase its revenues through increased mobile application sales, and reduce customer churn. The Hadopp solution can consolidate all information requirements in a single environment, and enable reliable, ad hoc analysis and end user self-service.

This accelerates the delivery of critical business performance information to the point of need, in a timely enough fashion for that intelligence to be useful and actionable. The Hadoop solution is able to handle large volumes of data, be easily configurable by users, and provide graphics of the results, including the ability to drill down into the detail by way of dashboards.It's all automatic. Before, users would be sending emails and calls to chase the data.With Hadoop anyone across the whole business can have access to the information they need, and find it on their own.

Big Data in the enterprise should not live in a vacuum. It materializes from dozens of databases, applications, and external sources. It has to be ingested, transformed, enriched, analyzed, and ultimately shared back with the people who will use it to make decisions and serve customers. Hadoop can handle all types of data from disparate systems: structured, unstructured, log files, pictures, audio files, communications records, email– just about anything you can think of, regardless of its native format. Even when different types of data have been stored in unrelated systems, you can dump it all into your Hadoop cluster with no prior need for a schema. With network data volumes on the rise, it is imperative for telecom companies to keep a close watch on their networks to keep them functioning at peak performance, which is the key to retaining customers. Relying on data samples or aggregations isn't an option. Hadoop is the one super system that can scale to accommodate these data volumes at a reasonable cost.

The convergence of efficient wireless protocols, improved sensors, cheaper processors, and a bevy of startups and established companies developing the necessary management and application software has finally made the concept of the Internet of Things (IoT) mainstream. The number of Internet-connected devices surpassed the number of human beings on the planet in 2011, and by 2020, Internet-connected devices are expected to number between 26 billion and 50 billion.As with many new concepts, IoT's roots can be traced back to the Massachusetts Institute of Technology (MIT), from work at the Auto-ID Center.

Founded in 1999, this group was working in the field of networked radio frequency identification (RFID) and emerging sensing technologies. The labs consisted of seven research universities located across four continents. These institutions were chosen by the Auto-ID Center to design the architecture for IoT.The vast majority (80%+) of IoT connections will occur on unlicensed wireless frequencies due to cost and battery life advantages. Whereas cellular IoT connections are expected to grow at a nearly 20% CAGR for the next several years, various personal area network (PAN) wireless connections (Wi-Fi, Bluetooth, Zigbee) into M2M (machine-to-machine) end markets should grow closer to 30%. Aside from significant growth in PAN-based chipsets, some experts view the microcontroller and sensor opportunities as meaningfully positive with industry forecasts of single-digit growth potentially being too pessimistic.

In any case the IoT $ opportunity is not lost to Telcos and the senior execs are making noises now while others are forging ahead. Some Operators have already taken the lead in supporting such global service launches in early market categories such as automotive, health and consumer electronics. With the emergence of new products in adjacent categories such as healthcare, wearables and consumer electronics the importance of the ability to support large-scale global deployments is likely to accelerate.

According to Cisco IBSG there are several barriers, however, have the potential to slow the development of IoT. The three largest are the deployment of IPv6, power for sensors, and agreement on standards. It is important to note that while barriers and challenges exist, they are not insurmountable. Given the benefits of IoT, these issues will get worked out. It is only a matter of time.The challenge for operators is to find a business model that delivers value for customers and is profitable. A significant part of their challenge is determining what an IoT network architecture and business operation should look like. These dilemmas need to be resolved quickly because, within the next five years, serving the IoT market will become the critical mission for any communications service providers.As industries such as automotive, utilities, transport and logistics feel the competitive pressure of IoT, the scramble for partners to help them will accelerate. One of the principle capabilities these companies will seek of their partners will be their ability to deliver complex solutions quickly.

Telcos are busy assessing alternative options for new Low-Power Wide-Area (LPWA) networks, since the connectivity revenues alone from these networks are exceedingly low: typically USD2–3 per device per year, but below USD 1. With such low revenues on offer, Telcos are wondering whether they should invest in LPWA networks. GSMA believes that While connectivity will underpin the development of the Internet of Things, to avoid becoming commoditised, mobile operators must leverage their networks' potential to provide value added services and build what could become a US$422.6 billion industry. In the case of the overall market revenue of US$422.6 billion, the majority of these revenues are to be derived from the 'Service Wrap'.

The 'Service Wrap' comprises the service that the end customer pays for that relies on the underlying connectivity, and operators are investing in building new capabilities that improve their offering to IoT service propositions. Examples include horizontal capabilities such as remote provisioning of IoT devices, building platforms that allow for management of business rules, reporting, support for Application Programming Interfaces (APIs) and the management and presentation of data. Moreover, 'Big Data' analytics is set to become a key part of IoT services in the future, with operators increasingly looking at ways to analyse data from various sources and create new service lines.

Telefónica has launched a modular internet of things platform called Thinking Things, which consists of stackable modules for a variety of purposes.There will be many sensors, actuator modules and so on to come, but the first manifestation of the new platform is an "ambient kit pack" that includes a communications module with an embedded SIM, a module for measuring air temperature, humidity and ambient light, and a battery module that can be charged via microUSB (the battery modules,which can charge 1,000 "communications" per charge,) can themselves be stacked.

This will let users remotely control the temperature, lighting and humidity of their home or office, though that only applies to lights, heaters and humidifiers that are plugged in at the wall, rather than fixed units. That use case will probably also require the smart plug module that Telefónica will release early next year, allowing users to turn devices on and off, dim lights and measure energy usage. Telefónica has released modules for sensing presence, impact and audio, and notifying the user via LEDs. "This is a major step in Telefónica's journey into the internet of things," the company's director of industrial internet of things, Francisco Jariego has said. "Our aim is for Thinking Things Open to become an open ecosystem in which any object or device can be connected to the internet."

The IoT will increase the range of services, each requiring varying levels of bandwidth, mobility and latency. For example, services that are related to public safety or personal safety will generally require low latency, but not high bandwidth per se. alternatively, services that provide surveillance might also require high bandwidth. Due to the differing level of service demand, mobile networks may need the ability to identify the service which is generating traffic and meet its specific needs. For example, alert services related to public safety or personal health would require a higher priority compared to metering information, which is a normal monitoring activity.

For every Internet-connected PC or handset there will be 5-10 other types of devices sold with native Internet connectivity. These will include all manner of consumer electronics, machine tools, industrial equipment, cars, appliances, and a number of devices likely not yet invented. In the world of IoT, even cows will be connected. A special report in The Economist titled "Augmented Business" described how cows will be monitored .Sparked, a Dutch start-up company, implants sensors in the ears of cattle. This allows farmers to monitor cows' health and track their movements, ensuring a healthier, more plentiful supply of meat for people to consume. By the way on average, each cow generates about 200 megabytes of information a year. Bottom Line : Jump on the IoT bandwagon for the right reasons. Be prepared.. its not an trial initiative but a way of life !!

In developed and developing countries alike, market saturation in telecoms is limiting customer acquisitions and value added services have not been able to generate the same revenue as voice services.Cloud offers a unique opportunity to service providers that want to offer value added services like voice, video and collaboration on cloud platforms, but success will come only with simplicity and a recognition that the economics of the cloud are very different than traditional telco models. You maye be surprised to learn that cloud models have utilization patterns like airlines which means : they are capital intensive, time and context sensitive and as such supply and demand differentiation becomes critical to maximizing yields.

Telcos today have around 5% of the public cloud market, according to analysts' estimates. They could potentially increase their market share by going beyond the provision of connectivity to provide additional services, such as authentication, billing, systems integration and even professional IT services. While most telcos can't match the IT expertise of IBM or HP, they have some advantages, such as long-standing relationships with both large and small businesses, well-known brand names and extensive customer care facilities. Technologies are considered to have become "mainstream" once they have achieved 25% penetration. As cloud follows this same trajectory, with a slew of telcos, cable operators, data centre specialists and colocation providers entering the market, significant consolidation will be inevitable, since cloud economics are inextricably linked to scale.

Cloud Computing attracts investments and overseas businesses and provides a significant boost to e-government initiatives. A Cloud Readiness Index is a good idea since it can track a continent's cloud adoption progress . By mapping the conditions and criteria required for successful implementation and uptake, one can identify potential bottlenecks that could slow cloud computing adoption. This "Cloud Readiness Index" would analyze key criteria that impact the deployment and use of cloud computing technology across different countries/cities : such criteria might include : Regulatory conditions , Data protection policy , Broadband quality, Power grid quality, Internet filtering , Government prioritization and ICT policy etc.

Without a doubt countries with the most insightful, transparent and fair regulatory environments supported by a the highest political echelons will be the most successful in capitalizing on this new opportunity.With ICT for Development representing an area of high interest for the International Telecommunications Union, the World Bank, and other development agencies, it is inevitable that the services of cloud computing would be applied towards various areas of socio economic development. These area include e-education, e-health, e-commerce, e-governance, e-environment, and telecommuting.

On a Global basis terrestrial and wireless broadband networks are ramping up catalysed by the arrival of several submarine cables at various landing points along the continental coastlines. What may slow cloud uptake is lack of 4G Digital Dividend Spectrum and delays in terrestrial fibre without which Mobile Cloud Computing will remain a sluggish experience. Unreliable networks undermine the entire cloud concept and unfortunately most networks in Developing countries are battling with infrastructure bottlenecks and capex constraints.

Mobile operators can build their own dedicated, private cloud or they can outsource it to a third-party to host. Operators around the world are giving strong consideration to outsourcing due to the economies of scale that cloud service providers offer and the capability to shift CAPEX to OPEX to better manage costs. As providers of cloud services, telecom operators can manage connectivity, deliver cloud capabilities, and leverage network assets to enhance cloud offerings. Given their core competency, managing cloud connectivity appears to be the most natural value-adding activity.

Identity management and security also came through as strong themes and there is a natural role for telcos to play in the cloud Telcos already have a trusted billing relationship and hold personal customer information. Extending this capability to offer pre-population of forms, acting as an authentication broker on behalf of other services and integrating information about location and context through APIs would represent additional business and revenue generating opportunities. Another opportunity is driving productivity and efficiency gains for Enterprises, together with improved customer service and increased revenues : by allowing them to incorporate CSPs' communications and context-based capabilities directly into business applications such as Field Service Management.

Recent forecasts suggest that there will be up to 50 billion mobile connected "machines" over the coming years, including appliances, smart meters, security systems, healthcare devices and many others – all of which can benefit from network capabilities accessed on-demand from the mobile cloud. A global pioneer in M2M, Norway's Telenor Connexion is offering its M2M services through a dedicated platform that enables Telenor Connexion to focus on delivering highly responsive market offerings and developing differentiating value-adds, such as customisation services closely tied with connectivity. It serves M2M customers in automotive, fleet management, security, utilities and healthcare.

With a culture of aiming for five nines (99.999% uptime) reliability, telcos are well-suited to the delivery of cloud services dependent on continual connectivity. By leveraging their network assets, operators add value by exploiting user attributes such as profiles and activities, making cloud services relevant and meaningful to users and providing the linkage between the upstream and downstream components of two-sided business models. Telcos clearly have a pivotal role in the cloud value chain and Verizon Communications, Deutsche Telekom , SingTel , Etisalat and other telcos are moving aggressively into this market.

For example Telefónica was facing growing pressures on its revenues and profits. Meanwhile, companies in the critical field of small and medium-sized enterprises (SME) were requesting new services to strengthen their business capabilities.Telefónica gradually settled on software as a service (SaaS) as the potentially optimum way to meet the needs of these customers quickly while keeping costs to the minimum. The two key cloud platform technologies to satisfy Telefónica' s demands regarding its SaaS solution were "Aggregation Skills" and "Multi-Tenancy ".The aggregation skill includes not only technology but business processors to aggregate and bring applications to the platform. These make it possible to deploy new applications very quickly and even globally if required thereby enabling the Telco to achieve benefits from the economies of scale . Multi-tenancy enables telco systems to accept upstream customers as operator-like entities, which could inject their own business rules into the system, use its development APIs, and run their product management independently.

Sometimes becoming a cloud service brokerage and bundling those services with their existing traditional telecom services confers certain advantages. One of the main benefits is service velocity, as XaaS services (Software-as-a-Service, Infrastructure-as-a-Service, etc.) can be on-boarded for pricing, fulfillment and billing in weeks, as opposed to the typical months-to-years timeframe. In addition, the costs to implement the services are dramatically reduced as well. That's what Deutche Telekom did with their Business Market place cloud for SME market in Germany.

Just as telecom operators are promoting cloud as a change agent for business, they too can benefit from its adoption. With operators seeking to transform themselves from their legacy environments and mindsets, adoption of cloud services can lead to efficiency gains, operational flexibility, and substantial cost savings. Perhaps that is where Telcos should focus on to start with while the spectrum , fttx , Government support and other bottlenecks are sorted out.Some key questions that must be answered while developing Telco Cloud strategies include : How can we best monetise the new service mix? How can the sort of margins available from voice services be realised for data? How can our business model evolve towards the sort of media retail models proposed by so many industry commentators? France Telecom Orange believe that in the initial phase Cloud services may not generate that much direct revenue but it reduces churn and that is good enough benefit to start with.

A ubiquitous mobile cloud will benefit the Global telecoms industry as a whole, by making it much more attractive for application service providers to create new services, or to enrich existing services, that use capabilities, information and intelligence provided by mobile and fixed telecoms operators. Eliminating fragmentation will result in a much larger addressable market of ASPs, resulting in increased service innovation, customer satisfaction, and new revenue sources for the industry as a whole, and consequently for individual operators.

Deutsche Telekom has now kicked off the next stage of digitization in the German and European economies by launching the Open Telekom Cloud on Monday, a public cloud combining flexible IT resources with service and strict German data protection regulations. Deutsche Telekom now provides a comprehensive cloud portfolio all under one roof: private and public cloud, software solutions including integration in companies' existing IT infrastructure.

"We are adding a new, decisive cloud offering to our existing portfolio of private cloud services that can be reached easily from the public Internet", says Deutsche Telekom CEO Tim Höttges at Cebit in Hanover. "For our customers, whether major corporations or SMEs, this is an important new service for their digitization - and an essential milestone for us in our ambition to be the No. 1 provider of cloud services for business customers in Europe." the public Internet", says Deutsche Telekom CEO Tim Höttges at Cebit in Hanover. "For our customers, whether major corporations or SMEs, this is an important new service for their digitization - and an essential milestone for us in our ambition to be the No. 1 provider of cloud services for business customers in Europe."

Thanks to the Open Telekom Cloud, Deutsche Telekom is entering a market segment that up to now has been dominated by its U.S. competitors. The technology firm Huawei is contributing the hardware and solution know-how, and T-Systems, Deutsche Telekom's business customer arm, is providing the data center, network, and cloud operation and management. Deutsche Telekom's multi-award-winning network will provide reliable availability and T Systems' certified German cloud data centers will ensure top quality and maximum security.

The Open Telekom Cloud will be set up in Europe's most cutting-edge data center, located in Biere, Saxony-Anhalt. Consequently, any data processed will be subject to Germany's strict data protection law. The Biere data center, and its twin center in Magdeburg, host almost all of Deutsche Telekom's ecosystem of technology and software partners. Deutsche Telekom's "House of Clouds" thus literally offers short distances to link one application to another, one cloud to another. Thanks to the extensive experience of T-Systems' cloud experts, data and whole application environments can be transferred easily from the public cloud to an even better protected private cloud.

Deutsche Telekom aims to double its revenue from cloud-based services for business customers by the end of 2018. Last year, revenue from cloud services rose by a significant double-digit figure at T-Systems alone. Up to now, customers have mainly used the specially secured private cloud. Deutsche Telekom and its subsidiaryT-Systems have been offering secure end-to-end cloud solutions for companies of all sizes since 2005 – from consulting, implementation, billing and customer service through to maintenance. Deutsche Telekom's growing partner ecosystem includes solutions from Microsoft, SAP, Cisco, Salesforce, VMWare, Huawei, SugarCRM, and Informatica.

Developing country MNO's are notorious (among many other failings) for their high dependency on diesel to fuel their base stations. One would think that faced with falling voice ARPU and hypothetical additional data revenue , energy expenditures would top the list to reduce OPEX since oil prices will remain stubbornly high at + $ 100 pb. Reduction in fuel OPEX requires CAPEX because it implies purchasing more energy efficient equipment or switching to renewable energy power solutions. While solar and wind remain the most prominent green technologies used to power off grid base stations , SON is another technical innovation within the 3GPP standards to save on BTS energy consumption.

So what is SON and how does it save on energy consumption ? A self-organizing Network (SON) is an automation technology designed to make the planning, configuration, management, optimization and healing of mobile radio access networks simpler and faster. SON functionality and behavior has been defined and specified in generally accepted mobile industry recommendations produced by organizations such as 3GPP and the NGMN . The first technology making use of SON features is LTE, but the technology has also been retro-fitted to older radio access technologies such as UMTS since Telcos began to understand that to meet the rising demand for data, it could be more cost-effective for them to expand HSPA and HSPA+ high-speed data capacity on the existing 3G infrastructure in many locations. SON promises enhancements in network efficiency, reductions in CAPEX and OPEX, improvements in customer experience (with potential reductions in churn).

So what are some of the benefits of SON in your BTS topology ?? A SON delivers an intelligent network where base stations self-optimize their operational algorithms and parameters in response to changes in network, traffic and environmental conditions. With operational intelligence at the access point, a SON can collect live network and call data, process it in real time, and either preview the changes or automatically deploy them live. SON offers offline planning capabilities for rapidly modeling the optimization of several parameters, including cell list additions, handover, interference control, and QoS enforcement. 3GPP Rel-11 has defined two energy saving states for a cell with respect to energy saving namely: not Energy Saving state and energy Saving state. When a cell is in an energy saving state it may need neighboring cells to pick up the load. However, a cell in energy Saving state cannot cause coverage holes or create undue load on the surrounding cells. All traffic on that cell is expected to be drained to other overlaid/umbrella cells before any cell moves to energy Saving state.

It is an indisputable fact that the traffic load in mobile networks is very unevenly distributed both over time and over cells. Excessive waste of energy occurs in low traffic situations since the radio system is optimized for maximum load. The NGMN Alliance has measured the daily average traffic distribution for an urban scenario .There is no communication activity in the cell in the two hours between approximately 04:00 and 06:00, and traffic is lower than 20 percent in the seven hours between approximately 0:00 and 07:00. So subscribers use more communication services during the day and very few in the wee hours of the morning. Knowing such usage habits is useful to determine how resources should be allocated so that the maximum amount of power can be saved. With SON base stations the network's coverage and capacity can be optimized when SON base stations can dynamically alter parameters such as antenna tilt and reference power offsets to compensate for lapses in coverage and ensure adequate capacity where it's needed.

Drastic improvements can be achieved by adapting to the actual traffic demand in a mobile network. The solutions include automatically switching off unnecessary cells, modifying the radio topology, and reducing the radiated power with methods such as bandwidth shrinking and cell micro-sleep. The challenge is to maintain reliable service coverage and quality of service (QoS) in the related area, while simultaneously consuming the lowest energy. The self organizing network (SON) supports proper selection of the appropriate energy saving mechanism and automatic collaborative reconfiguration of cell parameters with the neighbour cells.Mobility features like handoffs from one cell to the next can be optimized in a SON when base stations can balance load traffic among contiguous cells

Most engineers know that the amplifier power supply uses 60 percent to 80 percent of the energy consumed by base stations .If the RF power amplifier (PA) works at full power when there is no traffic or the load is very low, then power is wasted. Fortunately PA voltage can be dynamically adjusted according to the traffic load and required output power. When the output power is relatively low, the voltage required by a power amplifier is set lower than its maximum output voltage. In this way, power amplification is improved when the traffic load is light, and power consumption is reduced.In conventional amplifiers this power is independent of the amplifier input signal, i.e., of the current traffic load.

The key approach to saving energy is to make the power consumption proportional to the traffic load, either by implementing a partial shut down of amplifiers or by employing enhanced power amplifiers. During power-off of the amplifiers, further power savings can be achieved by also switching off the baseband signal processing, and indirectly, in the AC/DC power conversion and in the cooling fans.From the energy consumption point of view, low loads should be avoided. Instead, two types of mechanisms can be applied to reduce idle and unused capacities. As a first step, all energy-consuming equipment should implement power-reduction mechanisms while in operational mode, adapting to the actual load (short -term strategies). Second, the traffic should be reshuffled to a smaller number of highly loaded sectors or processing entities, and the others should be switched-off (long-term strategies). However, the coverage and the quality of service must not be degraded.

When there is no traffic, power can be saved by adjusting the PA voltage. Power can also be saved in transmitting LTE OFDM subcarriers. OFDM symbols can be automatically turned off when there is no baseband data transmission. This reduces power in the PA. Turning off PA OFDM symbols is more efficient than keeping them turned on all the time, especially when there is light or no traffic. PAs, cells, and power supply can also be intelligently turned off in the same way as OFDM symbols. A basestation's handling of RACH (random-access channel) offers another optimization metric. Automatically setting up a SON base station's RACH config parameters such as the number of preambles on a packet and ramp-up power can reduce synchronization times, call setup times, and handover delays while improving other aspects of RACH performance.

SON base stations are able to automatically configure themselves from the moment they are first powered up and before they join a wireless network. Once power is supplied, the base station would configure its physical cell identity, including its Internet Protocol (IP) address, and it would authenticate its software and configuration data. Following the completion of these baseline tasks, the SON base station would initialize the configuration of its radio by setting up its relationships with its neighboring cells and compiling its neighbor list. Based on a number of predetermined operational criteria such as energy savings, range requirements, and interference conditions, a SON base station will begin the self-optimization process once its initial configuration has been completed and it has joined the network. One of the first optimization tasks it will undertake will be to dynamically prune and select the base stations that are on its neighbors list.And this is where it really counts !!! A SON base station can save up to 40% of power usually consumed.

The combination of dynamic PA voltage adjustment and intelligent turning off of OFDM symbols is unique in the industry and can save about 32% of power consumption. Suppose the average power consumed by each base station is 1500 W (configured with three sectors). A single station can save up to 5200 kWh each year. This means more than 5.2 million kWh can be saved for a network with 1000 base stations each year, which is a saving of 1730 tons of standard coal and a reduction of 4500 tons of carbon dioxide a year. If base station power consumption is reduced, then less auxiliary power and heat-dissipation devices are required and less network OAM is necessary. The power needed for these devices is also reduced. New energy sources such as solar, wind, and bioenergy can be used in conjunction with these innovative energy-saving technologies. In this way, network energy consumption can be cut by + 50%.

The recent deployment of LTE to address the growing data capacity crunch, has highlighted the need and value of self-organizing capabilities within the network that permits reductions in operational expenses (OPEX) during deployment as well as during continuing operations. Self-optimizing capabilities in the network will lead to higher end user Quality of Experience (QoE) and reduced churn, thus allowing for overall improved network performance. Self-Organizing Networks (SON) improve network performance, but in no way replace the wireless industry's important need for more spectrum to meet the rising mobile data demands from subscribers !!

SDN can reduce provisioning time from weeks to seconds, but internal IT and networking teams should not expect their deployments to go as quickly as that. Mobile operators that are serious about investing in restructuring their network infrastructure will need to plan carefully. The number of vendors in the SDN market continues to increase, and operators must take the time to choose those that best accommodate their needs. With the advent of SDN, the network can be dynamically programmed in a highly granular fashion, through the app itself. SDN/NFV technologies realize network services on shared standard hardware, allowing faster and easier modification of network configurations such as capacity and geographical location. By using SDN/NFV technologies, networks can be deployed for each individual service and optimized based on particular latency, bandwidth, safety and security needs.

--♠--

Telco Digital # 3 : The Customer Agenda

Telco Digital # 3 :
The Customer Agenda

Agenda!

Telco Global Connect

DIGITAL Telcos are the rage although the definitions vary but the desired end result is the same. The most successful CSPs are implementing customer centric systems that have the insight and visibility required to ensure every customer promise is fulfilled efficiently.The right blend of technologies will help the Telcos earn the Digital soubriquet. Think Analytics , flexible BSS/OSS , XaaS models and Virtualisation. Telecoms should learn from other industries, those that have undergone disruption and continue to innovate. For example, retailers centralize design, architecture, and operations so that local stores can benefit from the latest fashion items. They can plan and order with fast turnaround times. Engaging with customers in multiple digital dimensions opens a whole new world of contexts in which the service provider can interact with their subscribers, be it offering devices on Social Media or placing roaming package promotions on etickets.

According to a recent Aspect Software study, almost 75 percent of consumers prefer to solve issues on their own—almost one out of three respondents noted they would rather talk to a toilet than a customer service representative. Mobile operators suffer from some of the worst levels of customer satisfaction in the world eventhough all the tools and platforms to offer superlative experience are available to them. What a disgrace !!

The annual WDS Loyalty Audit has revealed that ONLY 35 percent of customers are highly satisfied with the mobile operator. More worrying for carriers is that a quarter of subscribers claim low satisfaction.The figures pose serious questions for operators, as feeling satisfied is intertwined with a customer's intent to repurchase. A unsatisfied customer is 8 times more likely to switch operators. Despite these disappointing figures, mobile operators still appear to be doing very little to understand their relationship with consumers or customer loyalty. This in turn means their loyalty programs and customer satisfaction schemes, vital to customer retention and solid business performance, remain outdated and fail to deliver.

Customer Experience Management is certainly one of the biggest current buzz words in the mobile arena. A lot has been said and written around CxM (sometimes called CEM) but still there is no unique industry consensus about what CEM actually includes. From managing the brand's perception across websites or street-shops to subscribers' actual appreciation about the services and applications they pay for, Customer Experience Management is claimed to be a part of every business process. A common misconception in the industry is that CxM is a replacement for CRM which simply is not correct.

As the industry moves into a growing market of digital services built on infrastructures that enable fast development and deployment of new services, the service portfolio itself is not sufficient to establish a lasting differential in the market place. Such a differential is quickly eroded by competitive service providers. Having tried to differentiate through technology and 'clever' pricing models and found the strategy to be short lived, service providers are realising that a more solid differentiation can be gained through managing the customer experience. This does not just mean delivering service that meets the customers' expectations but that all aspects of its business must support the concept of a superior customer experience.

Many countries have a penetration rate above 100%. In such a competitive market, churn has become the major concern of operators who have changed their priorities from customer acquisition to customer retention. Operators' financial reports show that, for a medium sized operator the average cost of retaining an existing subscriber amounts to tens of dollars per subscriber per year.Comparing this with "Cost Of Acquisition" for new subscribers (COA), often worth several hundreds of dollars per new subscriber, operators are inclined to pamper their existing subscribers, especially those generating higher Average Revenue Per User (ARPU). In this context it is key for operators to understand user expectations and adapt mobile data plans to their needs.

A successful transformation into the CxM world can only be achieved by building on top of good CRM processes and practices. CEM takes us a step closer to achieving improved customer satisfaction. Instead of asking the question, "This is what we are doing, how well are we doing?" which is a CRM approach, CEM asks, "What is important to you, and how well are we doing?. CxM is aimed at turning customers into fans by seeing the world through their own eyes. In the May 2012 issue of Telecom Buzz, published by

MobileComm, the article "Customer Experience Management: The Next 'Buzzword'" declares the four pillars of telecom CEM ...no.... not nuclear physics or Astroflight.... but simply :

- network experience (includes coverage, signal quality, speed)
- commercial experience (includes billing, payment)
- product experience (includes telecom products such as handsets, VAS)
- service experience (includes after-sales service , customer queries)

In the social networking age you would already know the crucial role of social media and mobile marketing are the new building blocks when developing your future-proof CxM strategy. Increasingly Telcos are looking to looking to the social networking sites to provide valuable feedback on what the customer is experiencing. Twitter and Facebook provide rapid indicators on when things are going wrong. Systems that automatically monitor key social networking sites must be deployed to flag to the Service Management Centre when traffic increases. Often this is the first sign that a service is failing or a new service does not work the way that it should. And you can do something about it.... check this out ::

Delivering an effective Customer Experience Management requires a coordinated program across the entire organisation and is best achieved by adopting a maturity framework similar to the Capability Maturity Model Integration (CMMI) framework. CMMI is a proven process improvement approach whose goal is to help organizations improve their performance. The TM Forum is developing a maturity model for the implementation of CEM. This CEM model, as with CMMI, is a five stage model that guides the service provider on a journey to a fully implemented and controlled CEM environment.

The measurement of Customer Experience is based on measuring the extent to which the customer's needs are satisfied using customer/user centric measures such as: + Would advocate (e.g. churn and loyalty indicators) + Would recommend (e.g. Net Promoter Score) + Would Buy again + Product availability + Product usability.Having the right tools and OSS / BSS environments in place to support CEM is absolutely critical to achieving the end goal. As such establishing an early dialogue with tools suppliers (internal and external) has to be a priority in the early days of the program if for no other reason than the lead times for delivering and integrating the necessary solutions.

Bear in mind having CEM people without equipping them with supporting tools will lead to frustration and will feed the 'naysayers' with ammunition to criticize or undermine the CxM program. CEM is likely to introduce new working practices which may to some, seem unnecessary and a hindrance to rolling out new digital services quickly. Without strong governance the program will become disjointed with different parts of the organisation going their own way instead of a single 'joined up' approach to delivering a good customer experience.

Network management plays a major role in an improved CxM. By avoiding network congestion and poor performance, telecom operators improve the quality-of-service (QoS) level, which eventually can reduce churn to a great extent and increase customer satisfaction with an operator. However, an improved network QoS represents only one of the factors that influence the overall "customer experience" equation. Strategies for reducing churn need to take place at every step of the customer life cycle. If marketing communicates something that is not supported by product quality, network infrastructure, billing processes, or customer care/service teams, the relationship worsens until, ultimately, the customer moves away.

Recently Telefonica implemented a suite of CxM tools and platforms to improve the end-to-end customer experience across mobile data, mobile voice, IPTV, high-speed Internet, cable, satellite and voice services. The platforms and tools enable their customers to troubleshoot and manage their digital experiences through devices such as mobile phones, laptops and IP set-top boxes, via dedicated web portal and apps.Telefonica's approach towards managing the customer experience embraces a multitude of critical success factors including customer surveys, social media activity, contact centre stats and service specific data.

Recognising that the CEM view is a complex but profitable undertaking if you get it right , forward looking Telcos such as Telefonica have developed an OSS/BSS environment that enables them to display disparate customer data in one single 'vital signs' view.From this single view Telefonica are able to calculate various Customer Satisfaction Index values which they can then use to drive their customer centric quality improvement programs.

For the telcos to remain competitive an overarching customer-experience strategy ultimately makes more business sense.With growing pools of data (both structured and unstructured), gaining individual customer insights and coming up with products and services that suit them would require a sophisticated level of business intelligence such as CxM, which would deliver performance analytics to all management levels. Telcos that continue to ignore the dire need of having an effective CEM strategy supported by appropriate tool sets might eventually find themselves thrown out of the race...so good riddance and the Telco industry will be all the better for it !!

The Telenor Digital Unit has the challenging remit of developing services that will guide Telenor towards a future as an internet telco.Telenor Digital creates globally scalable solutions within next-generation communication services, cloud services, e-commerce, and the "Internet of Everything". Telenor Digital also enables global distribution of its own and third-party services and support new ventures within digital entrepreneurship.

Telenor believes that key to building a successful digital service is making it easy to use. Operators are ideally placed to securely remove obstacles to logging into an app or a web-page. By succeeding in the digital identity space, mobile operators will become a more visible part of consumers' everyday digital life. In response to this challenge the Telenor Global Backend is a common cloud¬-based infrastructure was developed to provide a global, shared system for giving Telenor's customers access to Internet service. This implies providing easy Sign Up and Log In - as well as frictionless payment. As of today - nine out of 13 business units can offer payment of digital goods through Global Backend - e.g. on Google Play. In addition, six out of 13 Telenor BUs can offer services bundled together with a mobile subscription to their customers through the Global Backend infrastructure.

This also means that Telenor becomes more attractive towards partners because they have one integration point through the Global Backend, through which they can reach

172 million customers – instead of approaching 13 Business Units individually. At the heart of the global backend business model, is the customer ID – the unique key which all customer data is gathered around. The Connect ID is Telenor's global solution to authenticate end-users. In practice – it's a solution for signing up and logging into a service. In short, Connect ID offers easy access to all their services.

Malaysia's YTL, which has built out a nationwide 4G network to help close the country's digital divide, has also created what it says is the world's first national education cloud that is deployed in all 10,080 primary and secondary schools in coordination with the government's education transformation blueprint. With its 4G network, the company can offer anytime, anywhere learning. They have put in extra measures to create an VLAN-over-4G architecture to protect the children while they are connected. All it is backboned on a cloud-based learning platform, called Frog VLE (virtual learning environment), to create an intuitive experience for students, teachers and parents.

Launched in November 2010, YTL Communications was a pioneer in the delivery of mobile telephony on a mobile broadband network. Subscribers get both Internet and Voice in a single plan with voice roaming being free. Their 4G network is SIM-less and runs on a user ID that comes with its own mobile number. Subscribers can log on to multiple devices, all at the same time. Their Yes 4G service is a mobile internet platform that facilitates rapid and continuous innovation offering a ground breaking architecture that gives users unmatched performance, convenience and cost savings by bringing together mobile broadband, mobile telephony (voice and SMS) and cloud-based services across multiple devices, all in a single Yes account !!

YTL is the connectivity partner for the Malaysian Govt1BestariNet project which seeks to transform the education arena and establish Malaysia as a model of excellence in integrated, Internet-enabled learning. This to be achieved by providing all state schools with YTL 4G connectivity to Frog's Virtual Learning Environment (Frog VLE), which offers exceptional control enabling teachers, admin staff and even pupils to fully embed their learning platform into the school's working practices and tailor it to the needs of their school.

The Frog Virtual Learning Environment (VLE) is a web-based learning system that replicates real-world learning by integrating virtual equivalents of conventional concepts of education. For example, teachers can assign lessons, tests, and marks virtually, while students can submit homework and view their marks through the VLE. Parents can view school news and important documents while school administrators can organise their school calendars and disseminate school notices via the Internet.Combining high-speed 4G internet access, a world class learning platform and access to 'best-in-class' resources and technology, with YTL , Malaysia is the first country in the world to bring its entire education community together on a single converged network designed specifically to meet the needs of teaching and learning !!

Across many industries, from entertainment to banking, from health to e-government, services and processes are becoming both more digital and more mobile, yielding efficiency and convenience benefits for individuals and businesses alike. However, consumers want to be able to access these services securely, shielded by robust privacy safeguards and strong data protection.One of the GSMA's top four priority programmes, the Personal Data initiative has developed Mobile Connect, a fast and secure login system that enables individuals to access their online accounts with just a single click or, where appropriate, automatically. Mobile Connect can provide different levels of security, ranging from low-level website access to highly-secure bank-grade authentication. Mobile Connect promises to make passwords a thing of the past. To use the service, individuals subscribing to a participating operator simply need to click on a website's Mobile Connect button.

For the uninitiated the GSMA Mobile ID is a PKI-based secure authentication service that enables users of business applications to access secure accounts, platforms, applications and cloud services in a single, unified mechanism. The service both simplifies the user experience and protects the individual's identity as they interact in the digital world. Mobile ID is already preinstalled on the SIM card as a SIM toolkit (STK) applet, which can only be accessed by the mobile provider "over the air" (OTA) via the correct ID key. As a result, Mobile ID works on all mobile devices that meet the GSM standard (GSM 11.14 or 3GPP 31.111), regardless of the operating system.

For digital service providers such as Telenor, Mobile Connect will deliver the optimum balance between convenience, security and privacy. By enabling consumers to log in quickly and easily, it improves the customer experience and reduces the likelihood that authentication issues will lead to them abandoning transactions, while minimising fraud and errors. In this way, Mobile Connect increases customer confidence and strengthens loyalty, while enabling fast engagement and enhancing the digital service provider's brand.

Shared data is a success story for savvy mobile operators. In today's multidevice ownership market having a pool of data that different customers, with many mobile devices, can share is proving a win-win for operators and customers. In order to work correctly and deliver the desired results there are some basic building blocks that shared data plans need to be incorporated into the Telco OSS/BSS .

These technical foundations include Real-Time Usage Tracking and Balance Management ; Shared Allowance Profile Management; On Device Plan Self Management ; Adding parental Controls and Offer Management via a catalogue. As most of us already know that the central importance of real-time capabilities in OSS/BSS permeates all aspects of operational, network and business management. Since many new services demand real-time support, operators must either transition to real-time OSS/BSS capabilities or forego these revenues. According to Current Analysis Operators will be "using virtualization to drive innovative service creation, especially the creation of services and apps that require time to market intervals of only days, even hours.

There is a wide range of applications of shared data. These include multi device plans, family plans, group plans and business plans. There are two foundations of success for shared plans. First is giving the customer control in setting the plan and second is real-time balance management. With many different users and devices using the same data pool, it's vital that balances are managed in real-time, and that customers are fully in control of their usage and have real-time visibility of their charges. A look at the Q2 2014 results of some of the main innovators of shared data underlines the benefits that operators are realizing.

AT&T's Mobile Share shared data plans, now represent more than 41 million connections, with the number of Mobile Share accounts more than tripling year-over-year to reach 14.6 million, with an average of about three devices per account. 49% of Mobile Share accounts had 10 GB or larger data plans, up from 25% in Q2 2013. Mobile Share has helped drive year on year increase of 20% in wireless data billings. Year on year Verizon has increased revenue per account by 4.7% and growing the percentage of accounts on More Everything plans from 36% to 50%.AT&T's plans cover up to 10 devices per Mobile Share Plan and range from 300MB to 50GB of data to share. Plans included unlimited voice and SMS, and AT&T also offers 50GB of free cloud storage with AT&T Locker, which is marketed as a secure and safe place for customers to store their photos. When it comes to devices to add as well as smartphones, gaming devices, tablets and so on.

Operators want customers to use tablets on their networks. Tablets are driving subscriber growth and operators are rolling out innovative offers to get customers buying and using cellular enabled tablets. As an example of contributions tablets can make to an operator's results, of the 1.4M retail net customers Verizon added in Q2 2014, 304,000 were postpaid phone net additions and the remaining 1.15 million were postpaid tablet subscribers. Mobility is a key driver for mobile data connectivity on tablets. This is particularly the case in emerging markets, where a higher proportion of tablet users than in more-mature economies report using their devices outside the home and on the move.

While the global tablet market is stabilising (IDC forecast 2014 worldwide tablet shipments of 233.1M units: a 6.5% year on year growth rate, after several years of double and treble digit growth), the number of cellular enabled tablets is on the rise. In Asia-Pacific, according to IDC, 25% of total tablet units shipped in the region have built-in option of voice calling over cellular networks (a 60% year on year growth). Shipments of tablet PCs to South Africa increased 107.1% year on year in the final quarter of 2013 to total 513,000 units .

In August 2014 T-Mobile also launched a tablet promotion for its customers on their Simple Choice plan. T-Mobile are matching the amount of data of a customer's smartphone to their tablet for $10 a month (up to a limit of 5GB). This level of pricing shows how eager operators are to get customers using tablets on their networks. T-Mobile's messaging pushes ' no overage charges' (they throttle speeds) in a drive to increase cellular tablet usage.Consumers increasingly watch TV and video on tablets. This will not be lost on operators who are looking to offer entertainment services—such as LTE broadcast and roll out TV / video partnerships. Getting tablet subscribers on board now may help ease the launch of these services as operators look to offer entertainment bundles to existing tablet subscribers.

One of the pioneers of shared data plans, Bell offers 'Family Shareable' and 'Personal Shareable' options. Customers can connect up to 10 devices or family members to the share plan, and also offer Mobile TV as a plan add on. The Mobile TV add on enables customers to watch over 30 live and 14 on demand TV channels for $5/ month per device for 10 hours. Bell's $50 and $60 share plan includes the Mobile TV add-on free for 3 months.

Telstra Australia launched their "shared" data plans for consumer plans in 2013. The plans are pretty simple once you get your head around it. Basically, it's available as a bolt-on option with Telstra's Every Day Connect consumer-level plans. It's not included for free, however: you'll be paying for the privilege. Every Day Connect plans come with one SIM card by default, but you can have up to three SIM cards connected to the one plan for data sharing with your SIM-enabled tablet. For example, say you have a phone with Telstra and an additional SIM-enabled tablet. The $60 Every Day Connect Plan gets you $600 of calls, unlimited text and MMS plus 1GB of data. To activate data sharing, you'll pay $10 for the Every Day Connect Data Share Plan, then an additional $10 for the additional SIM card for a final cost of $80 per month.

Sprint's Family Share Pack allows customers to connect up to ten devices or family members to a shared data bundle. Data bundles start at 600 MB and run all the way to 60 GB along with free unlimited voice and text. As well as their smartphones, customers have the option of connecting tablets ($10) and other mobile broadband devices ($20) to the plan. Where this offer is particularly interesting is in relation to new customer retention, Sprint will pay termination charges of up to $350 for customers porting over to this plan from another operator and also waive any access fees. This new plans has a limited running time, finishing at the start of 2016, which suggests Sprint are looking to entice new customers to sign up as well as make their current customer base "stickier".

In the UK, Vodafone is offering their own shared plan called "Red+". The plan aims to allow up to nine separate SIMs to connect to one "group leader plan". The group leader signs up to a data plan with an allowance of either 2GB, 4GB, 7GB, 10GB or 13GB and then defines how much of this allowance each member should receive. The plan is aimed primarily at families where the group leader would be one of the parents. Red+ plans allow the leader to both cap or expand the usage of each member, which is perfect for a parent whose teenager is a "data hog", and also offers free calls and texts to all members within the group. The really interesting part of this offer though is the fact that not only are there data notification alerts at 80 and 100%, but also you can't go over your allowance, unless you add extra data, meaning the subscriber is always in control.

UK 4G operator EE promotes flexibility in their 4GEE share plans, by enabling customers to add people to their plans at any time they want. The message is that if a friend (potential share plan member) is in the middle of a contract (presumably with a competitor) then the customer can add them when that contract finishes. This is a good example of using share plans to attract new customers. Starting at 250MB for $15 going up to 100GB for $750 Verizon's More Everything data share plan also provides unlimited talk and text offers. It also is offering a range of add-ons as standard – e.g. the plans come with 25 GB cloud storage and American Football (NFL) app offering NFL content and live games, an educational tools app, as well as the ability for customers to use their smartphones as a personal hotspot to get wi-fi enabled devices online.

There is no doubt that in the face of declining revenue from voice and messaging services, operators look for ways to monetise on data services and for pricing models that encourage customers to stay with their provider. Offering customers the option to sign up several devices with one subscription is not only attracting subscribers to sign up more devices (for example, by acquiring a connected tablet rather than a Wi-Fi only tablet), but also provides customers with an incentive to stay with one provider, if a single subscription provides cost savings, better matches their data consumption behaviour and facilitates the billing process.

When moving into new strategic verticals, Digital Telco have various go-to-market strategies in front of them. These strategies often range from "dipping in the toe" by partnering with companies that have vertical expertise to going the "whole hog" and building the expertise internally. When Australian incumbent Telstra decided that eHealth was going to be a strategic vertical, they resolutely jumped all in. Telstra created a stand-alone business unit, Telstra Health and hired a leading Healthcare professional, Shane Solomon to build it from the ground up. Telstra Health's mission is all about bringing the benefits of the digital revolution to health providers, practitioners and – most of all – its patients.Telstra's eHealth plan is to connect patients, healthcare workers, hospitals, pharmacies, government and health funds to build a safer, more convenient way of managing health. Telstra Health has mapped out various segments and is offering a tailored solution and product set to each.

Telstra Health offers some of the leading applications and solutions. These include Clinical Workbench, powered by Verdi; patient records aggregated in real time, and accessed on an iPhone or iPad; the suite of Fred applications, to help pharmacies manage, automate and dispense; and solutions for aged and community care workers, and GPs to assist in managing their day to day operations. Backed by Telstra's considerable expertise at software integration, experience and commitment, these applications provide a smoother, more efficient way of working – and a better experience for patients.Patients have never been more empowered. They do their banking and pay bills whenever they want. Order products and services online. Watch media as it happens live, half way around the world. Even report the news, themselves. So, why should they have to wait to get the health care they want and need? They don't have to. With Telstra Health services, patients can organise a time to see a GP, dentist or specialist when it suits them. And do it anytime of the day or night. They can talk to a medical expert over the phone or via video. All from home.

Hospitals are overburdened and there are long wait lists. As the population gets older, more people need care – and many require more complex care that must be coordinated across multiple providers. This task is time consuming and falls often to family members, who must piece together the puzzle as they juggle multiple responsibilities at home and work. Unless one can find a way to reduce admissions to hospital and aged care facilities,the health system – and many Australian families – will reach breaking point. To do this, we need to find smarter ways to coordinate care for those with high risk conditions. Telestra were convinced by the need to deliver care more proactively in home settings so that we can increase capacity in our care facilities.

Telstra Health, see an opportunity to remedy this by connecting eHealth technologies already available. For example, home monitoring devices can provide diagnostic information in near – real time with alerts so healthcare professionals can follow up with a call or a visit for patients whose vital signs are a cause for concern. For example,when blood pressure increases or there's evidence that medication hasn't been taken.By connecting this information to the applications used by GPs and nurse care co-ordinators, home healthcare can be organised quickly and efficiently, avoiding emergency visits or premature entry to residential aged care. It's already working in other parts of the world. The Ontario Telemedicine Network found that home health monitoring, combined with remote care coordination, lead to a 70% reduction in emergency department visits and a 60% reduction in hospitalisations.The ability to analyse 'big data' on hospital admissions can help identify who is at high risk of re-hospitalisation, and target health services to the people who need them most.

Giffgaff is a mobile virtual network operator (MVNO) : it uses the network infrastructure built by its parent company, O2. It may share the same technical back end as O2, but there the similarity ends. Not only has Giffgaff never run a call centre, all queries are answered online by its own members. The company's entire business model rests on its community: members not only help new customers, they recruit new "Gaffers" to the network, and come up with strategies to grow the business. To date, 14,000 ideas have been submitted to Giffgaff, which has implemented around 10pc of them, once duplicates have been removed.

Giffgaff has community message boards where you post a problem and the gamified experts will help you figure it out. In return, you help other people figure out their problems. What do you get out of this? Well two things really; One, you get resolution of problems by people that have actually experienced the problems (and hopefully the solution) and secondly it should result in a much cheaper mobile service, not having to pay for all those customer care agents.

Along with its unique structure, GiffGaff is also unique in its package offering as the network offers handsets on finance but doesn't offer pay monthly contracts. All GiffGaff phones are unlocked as well, meaning complete flexibility in the way you use your phone. Overall, GiffGaff is certainly a unique network and it definitely has a unique proposition. By offering contracts (and PAYG handsets) at cheaper than its rivals with less stringent rules, the network can appeal to all users who are disillusioned with their existing network.The unique finance offering will definitely appeal to customers who have difficulties getting accepted for traditional pay monthly contracts and in some cases at least, the finance option can still work out cheaper than buying the handset outright from the manufacturer.

A new survey by consumer watchdog Which? has found that mobile virtual network operators (MVNOs) like Tesco Mobile and Giffgaff have highest customer satisfaction than the big networks. Which? asked more than 4,000 people in the UK to rate their mobile provider in different categories, including value for money, and the ease of getting in touch with the customer service.It found that Giffgaff got the biggest thumbs up from users, with a customer satisfaction score of 79% – partly perhaps because it doesn't have a customer service department to contact, relying instead on community-based support. In second place was Asda Mobile (72%), followed by Tesco Mobile (70%) rounding out the top three.

Sometimes its not what you do but how you do it for your customers that entitles you to the Digital Telco soubriquet. T Mobile USA has done things with their Un Carrier Strategy that has wowed its customers , beefed up its income and bamboozled its competitors (Verizon , ATT). " Un-carrier" is a marketing campaign that debuted in March 2013, where the company introduced a new streamlined plan structure for new customers which drops contracts, subsidized phones, overage fees for data, and early termination fees.

Ever since John Legere took the reins as CEO in September 2012, T-Mobile has focused almost exclusively on doing things other carriers wouldn't, bringing about a series of consumer-friendly "Uncarrier" features that created free music streaming, data rollover, and a continent-wide phone service. T-Mobile's also asserted that by the end of 2015, it's LTE coverage will be much better. And those new ideas and LTE promises all seems to be working.

When T-Mobile started this whole Uncarrier thing, it began with the introduction of a Simple Choice phone plan, meaning free international data and texting when you signed. Almost two years later, T-Mobile is upgrading that original phone plan by expanding coverage to Mexico and Canada. It's called "Mobile Without Borders" and this is how it works: When you cross the border, whether north or south, your phone will work just like it does back in the states. That means calling, texting, and data, but also all the other T-Mobile Uncarrier stuff they've introduced over the past two years, will work in those countries.

But even if you're not a big traveler, making calls into those countries are now absolutely free as well. So basically, it's no longer a US plan but a North America plan for free.Data Stash," the company's new initiative lets you rollover your unused data month to month, rounded up to the nearest megabyte. f you don't use all the data you pay for one month, it will carrier over to the next month. To sweeten the deal, new and existing T-Mobile customers who qualify will also received 10 GB of data to start with. "This is probably the biggest thing we've ever done since Uncarrier 1.0," Legere.

T Mobile offers you to stream music for free (as in it won't count against your data limit) from most of the major music services. This includes Spotify, Pandora, Slacker, Rhapsody, iHeartRadio, and iTunes Radio, plus Samsung's Milk and Beatport when it launches.Not only will music not count against your data bucket, but it will continue to stream at high speed and full quality If you've already reached your data limit. This is absolutely fantastic. Basically every time I've gone over my data limit it's been because I was streaming music too much. Having to never worry about that is a big deal.

"Overall, we've done nine 'Uncarrier moves.' An Uncarrier move is a move intended to change a stupid, broken arrogant industry, and it's intended to be permanent, and we want everybody to change [to] no contracts, anytime upgrades and international data roaming,"

(Legere – CEO T Mobile)

T-Mobile and its "Uncarrier" strategy are great. Selling people reasonable phone plans without locking them into ridiculous contracts is fantastic and consumer friendly. Some sceptics say that good ideas don't always make good business sense. That maybe but the point about the pressing need for customer centricity and thrilling expereince has been made.

Unless you live in a cave or on Mars, you should know that social is the way to go. Social media – blogging, online social networking, and micro-blogging – have become so pervasive that it is almost unthinkable for a business entity – at least those who want to remain relevant !! In telecom, social media have transformed not only business models but the very concept of customer service. Emerging markets have embraced social media with gusto. Both India and Brazil represent some of the most aggressive growth, where more than 90 percent of online survey respondents report having an account on a social networking site. The reasons for this social media explosion in the emerging markets can be attributed to the concentration of Generation Y and younger, the cultural emphasis on maintaining regular contact with friends and family, and the influx of mobile technologies.

The IBM Institute for Business Value surveyed more than 1,000 consumers worldwide to understand who is using social media, what sites they frequent and what drives them to engage with companies. What the results showed may come as a surprise to those companies that assume consumers are seeking them out to feel connected to their brand. In fact, consumers are far more interested in obtaining tangible value, suggesting businesses may be confusing their own desire for customer intimacy with consumers' motivations for engaging. For companies that have been taking a "build it and they will come" approach to social media, these consumer findings are a wake-up call that much more needs to be done if they want to attract more than the most devoted brand advocates.

Telefónica Europe wanted to develop a European-wide social media strategy to align social media use across each of the company's business units and to help staff use it to enhance the whole customer experience. Telefónica Europe started by evaluating the company's current use of social media across all O2 branded businesses. Not only did this help to determine the strategic aims of each business unit, but, more importantly, it helped to highlight where the business was currently realizing value from social media, and the quick wins and opportunities for improving customer experience in the long run. Once the audit was complete, the next step was to develop a consistent strategy for the use of social media across multiple European markets in support of Telefónica's commercial brand, O2, and its Brussels-based Public Affairs department. This included developing three-year aims and objectives for Telefónica Europe's social business strategy; the whole process also took into account each local market's needs and conditions.

Salesforce.com was selected as the platform which allows a fast delivery of the solution in the cloud and also provides an integrated module for social media communication called "Chatter" lifting Web 2.0 Technology into the Cloud. Due to the scalability and multi-tenancy of the Cloud solution, further CRM applications can be easily integrated.Telefonica are developing, testing and embedding social media programmes across Europe in different departments and teams including Customer Experience, Brand, Communications and Customer Contact Centres. Education and training for 29,000 employees was implemented to embed the social media strategy and standardised processes across the business to ensure best practice and maximum return from social media.

As a leading Telco with over 23 million customers, O2 (subsidiary of Telefonica) decided to embrace social channels, as it is increasingly seeing a change in customers' buying and service expectations—with a growing preference to use the online channel. O2 started a business transformation journey to a multi-product, multi-channel company whilst continuing profitable growth. Evidence of this transformation can be seen in some of the services such as the new online shopping experience for small office/home office businesses. A Chatter app clears the line for employee communication O2's in-store support staff—called "gurus"— to collaborate on issues and help customers immediately, often while they are still standing at the counter.

In the Telco industry, loyal customers are the key to success. Nobody knows that better than Sprint, rated #1 for customer satisfaction and the third-largest telecommunications provider in the United States. When Sprint wanted to use new social media platforms to share information across groups and manage relationships with business customers more efficiently they consolidated customer information, automated processes, and built apps to make it easy to share data with retail stores. Information on business customers from multiple CRM systems was consolidated into customer profiles in the social media platform : a one-stop-shop for our sales teams for business processes and customer intelligence. More than 6,000 employees now use a Cloud platform to track accounts, contacts, and opportunities for improved visibility and real-time analytics.Sales Reps can quickly identify the best opportunities driving new leads and reducing the time to close enterprise deals by 25%. And, a fully-integrated configure-price-quote system helps Sprint's sales teams quickly build complex pricing models using up-to-date account information.

Customer visibility is enhanced by a Chatter application, which helps reps share and information and collaborate on deals so they can address customer needs quickly and provide consistently great service. Using a custom Internet portal, retail staff can easily share business leads with sales, and no leads get dropped. In the future, retail personnel will be able to collaborate with customer service professionals to solve customer issues on the spot. Custom apps built on the social media platform automate processes including managing waitlists for hot new phones, scheduling corporate briefings, or tracking churn data, so reps can proactively reach out to customers in danger of defecting or counteract competitive promotions. Another app manages the discount approvals process and recaptures almost $70 million in unauthorized discounts each month.

The benefits are social media based CRM are deep provided it is implemented strategically. First, there is the social interaction itself, which can provide direct value to the business through revenue from social commerce and cost savings when used for customer care or research, for example. Plus, social networking enables rapid, viral distribution of offers and content that may reach beyond what could be done in traditional channels – all with endorsement from connections people trust. But that is just the beginning. Companies also can use social platforms to mine data for brand monitoring and valuable customer insights, which can spark innovations for improved services, products and customer experiences. In a constant cycle of listen-analyze-engage evolve, Telcos can optimize their social media programs to continually enhance their business.

Telco Digital # 4 :
The Financial Agenda

Telco Global Connect

The build out of today's 4G networks such as LTE requires a dramatic increase in computational resources to adequately deliver flexible telecommunications services to mobile subscribers. Yet business conditions also necessitate that new markets are approached incrementally. The challenge for telecom carriers is to reduce the cost of serving the first subscriber in small or cost-sensitive markets. The primary challenge in serving small LTE subscriber bases is that traditional core network architectures require high capital expenditures just to serve the first subscriber.

Networks, whether entry-level or full-scale, are traditionally built using separate network elements for each of several different functions. And most network elements have been deployed with a pair of carrier-grade servers to achieve redundancy with an active and a standby configuration. Thus, a new network with 10 network elements requires 20 servers just to provide service to the first subscriber. Furthermore, because the network is designed to eventually support a large population of subscribers, the servers would remain underutilized until the subscriber base grows to the expected population. The ROI for small and emerging markets has therefore been limited by these high capital outlays. High operating costs for maintaining the servers and providing data center floor space, power, and cooling have also hindered new service opportunities.

The greatest opportunity for revenue growth for wireless broadband presents itself in the form of smaller markets with less than 50,000 subscribers, thereby lowering the cost dramatically to serve the first subscriber and the breakeven point in the Operator's business case . By dramatically lowering the cost to serve the first subscriber, new networks can be built on a campus or targeted community basis with new services tailored to the specific needs of these smaller, targeted markets.

According to Warren Buffet "The single most important decision in evaluating a business is pricing power..If you've got the power to raise prices without losing business to a competitor, you've got a very good business. And if you have to have a prayer session before raising the price by 10 percent, then you've got a terrible business." If this is true then it would make Telcos in the 4G world a lousy business since some Telco Execs have a prayer session on how to cut prices by + 25 % every 3 months and crow about it in the Media like pontificating politicians seeking votes during a presidential election.

The Boston Consulting Group Report (The Internet Economy in the G-20: The $4.2 Trillion Opportunity,) surveyed about 1,000 people in each of several G-20 nations on what "lifestyle habit" they would give up instead of the Internet for a year, including sex, alcohol, showers and cars. Most of the results for items like coffee, chocolate and fast food were steady with averages of 70-80 percent. Japan topped the list of citizens who would make the sacrifice, with 56 percent who would abstain from sex. Brazilians were the least likely to give up sex for the web access – only 12 percent surveyed would give it up. American and South Africans were most attached to their vehicles – only 10 percent each were willing to give up their cars for the Internet. Another interesting finding was the perceived value of the Internet versus its actual cost. For instance, Americans value the Internet at $3,000. According to BCG, it's value is actually $472 – an incredible markup in price based on perception.

So in light of the above : is there any hope for Telco execs who are intent on destroying the profitability of the Industry by engaging in vicious price battles that threatens the sustainability of the Industry ???? Thats where Yield Management can provide some insights on how best to price data. Robert Crandall, former Chairman and CEO of American Airlines, gave Yield Management its name and has called it "the single most important technical development in transportation management since we entered deregulation." Ditto for Telco De Regulation.

Believe it or not the telco and airline industries have much in common in terms of perishable inventory , large sunk low marginal costs and varying predictable demand volume. Yield management principles would indicate that, as the costs of using network capacity are minimal, it makes sense to use as much of it as possible when it is available and maximize revenue from it when it is in shorter supply. Unused bandwidth is lost forever.Yield management is the process of understanding, anticipating and influencing consumer behavior in order to maximize yield or profits from a fixed, perishable resource such as airline seats or hotel room reservations or advertising inventory or Internet bandwidth .

Its core concept is to provide the right service to the right customer at the right time for the right price.This process can result in price discrimination, where a firm charges customers consuming otherwise identical goods or services a different price for doing so.It is surprising if not shocking that Operators are utilizing on average only 35 to 40 percent of their network capacity. It takes some creativity to turn this huge dormant asset into profits.There are several scenarios in which yield management can be used to increase revenues.

1. Improving Market Segmentation and Pricing : Market segmentation enables enterprises to cater products and services, including pricing, targeted at the buyers in each segment. The basic idea, which makes segmentation an effective tool to increase an enterprise's bottom line, is to charge more for products targeted at customer segments with a higher willingness to pay.

2. Monetizing Unused Bandwidth : Telcos stand to increase profits through the creative monetization of their unused network bandwidth. Similar to the empty airline seat, network bandwidth is a perishable resource with a very low marginal cost, so filling the underutilized bandwidth with revenue-producing network traffic will have a direct, positive impact on the bottom line.

3. Dynamic Pricing : Market segmentation and associated segment pricing aims to maximize profits by fixing prices at levels that are optimal for targeted segments. Dynamic pricing encompasses adjusting the prices to changing market conditions and/or the status of the Telco's resources.

4. Increasing Profits at Peak Utilization : Although dynamic price adjustments can be used as an effective mechanism to off-load the network during peak usage, periods of peak utilization by definition are synonymous with periods of peak market demand and, as such, should be assessed for opportunities to increase revenues.

5. Reservations : The mobile network's bandwidth management functions enable Telcos to reserve bandwidth for specific customers in advance of their actual bandwidth usage. When a customer's reservation is in effect, the network ensures that the customer receives the requested bandwidth even if there are other customers competing for the same bandwidth.

6. Pricing flexibility : Real-time charging functionality provides Telcos with pricing flexibility at least on par with, if not better than that used by the airline industry. Telcos can charge based on the attributes of provided services, customer characteristics, context, network state, historical usage, etc.

Besides increasing revenue Yield management can also reduce the need to increase capacity, resulting in savings in investment for providers, which can be passed on to consumers as lower costs. The value of yield management for mobile broadband is an opportunity for service providers to manage the quality of a user's experience while achieving increased revenues in the context of the exponential CAPEX costs associated with servicing the global demand for mobile broadband services.

Telcos need to adopt a balanced approach on how to monetize 4G networks by a thorough analysis of various technical , financial and commercial strategies : the desired outcome is to minimize the cost of access coverage while maximizing subscriber capture. A balanced strategic plan is founded upon :

1. Effectively reengineer the broadband business model so as to reduce costs , manage data traffic , and develop a more sophisticated approach for pricing broadband access

2. Unlock new revenue streams to justify the enormous network investments over time in the context of key customer drivers and applications (cloud , M2M etc) that generate fast ROI based on understanding the needs of target markets.

3. Collaborate with OTT players since LTE's all-IP architecture will create a more open environment for Over The Top (OTT) applications which threaten to further commoditize the network.

4. Leverage the OSS/BSS to weaponise the CRM , Billing , Policy Control systems in order to ensure that all the data traffic is accounted for and billed to the correct entities.

What service providers need to do is to offer packages based on the service or application used that can be provisioned dynamically ,rather than on bandwidth allocation. Policy management tools play an important role here. By being able to offer management tools , the provider will be able to offer subscribers packages such as ' YouTube subscription' , or ' online gaming subscription' , or ' regular surfing and email subscription . Policy (PCRF) is the brains of a network , especially for LTE networks that must make many more real-time decisions to maintain network performance and adapt the network to the subscriber.

In a Media Research survey, respondents pointed overwhelmingly to smart devices, video, and cloud services as the devices and services most likely to drive demand for 4G. Fully 40 percent already have partnerships with content providers to assure them higher quality of service (QoS) on their networks. Enhanced enterprise LTE solutions, such as videoconferencing on-the-go and remote access to business applications, can drive data consumption. Verizon Wireless is one of many LTE operators that offers 4G mobility applications and solutions for SMEs and enterprise customers. A survey shows that 67% of US businesses using LTE believe that it has resulted in increased productivity.

In order to offer a more competitive service than the OTT players KT is leveraging CCC for ICT business. CCC is a kind of domain-specific cloud technology, based on virtualisation. By unifying the platform for radio and several application services into CCC, KT can provide cross-layer optimised services between applications and radio. For example, they can utilise user contexts such as user ID, traffic content, QoS, location, and the radio environment to provide the most suitable service to their customers.

There are some valuable options for addressing the data issue from a technical point of view, offload perhaps the most valuable amongst them. However, these are not all the weapons in an operator's arsenal. They can also look to manage the impact of traffic on their networks and their bottom lines by looking at different business model and pricing options. Operators can use the rich data experience of LTE to sell more data and develop new revenue streams. Video streaming providers such as Netflix alter the quality of video according to available bandwidth – so a 6-minute clip on LTE would consume 80MB compared with 27MB on 3G, thus driving usage. Operators are also bundling content with LTE or top-tier plans, enabling new revenue streams – for example, EE in the UK uses its film service (EE Film) to monetise data and receives sales commissions from video-on-demand provider FilmFlex.

On the revenue side, the bulk of revenue will be from 'downstream' subscription and pre-pay customers, and while helpful, that the near-term growth of new 'upstream' or wholesale / carrier services revenues alone would not be enough to cover the costs of capacity increases.Because LTE network latency is lower than 3G, operators can develop new revenue streams by selling bandwidth for wholesale services (such as utility and M2M services). Verizon is at the forefront of this with projects in sectors such as education.

MNOs can also experiment with bundling. Data sharing across devices is being offered, with the aim of monetising devices (such as tablets) otherwise lost to Wi-Fi. Tethering strategies are evolving, as operators try to monetise tethering by allowing it at as part of premium or top-tier plans. Fixed–mobile converged offerings are available and aimed at increasing revenue and reducing churn.There are many different possible solutions and different combinations of solutions will work at different times for different operators :

1. New air interfaces and spectrum will not be enough to on their own to cope with the continued rise in data traffic. Building more cells alone is not a solution, and it will be necessary to address costs and pricing

2. The challenge needs to be approached both from the network, through policy-based control including tiering and maybe traffic-shaping, backhaul optimisation, and offload through femto cells or WLAN, and from the business side with pricing, potential tiered offers and segmentation

3. Techniques have to be deployed to manage traffic to deliver customer experiences, particularly for cloud and TV services

4. Since no single method of addressing capacity issues provides a complete solution and therefore a combination of offload, traffic management and segmentation is recommended.

5. Mobile data optimization that includes content transformation is a crucial element in increasing the efficiency of data and video transport, by reducing the over-the-air payload on the RAN, and improving the subscriber experience with faster page loads and lower monthly data usage.

The companies that go to market with 4G services will have to be able to sustain them. The networks themselves will drive huge growth in data traffic. But changing business models also have the potential to explode transaction volumes. Not surprisingly, wholesale will be an important part of the 4G mix. The wholesale models most frequently cited are bulk access, machine-to-machine, and mobile virtual network operator.

However scalability and sustainability will also affect billing systems. CSPs will need to invest in the next generation of business-support systems (BSS) to manage customer-facing operations such as product, order, customer, and revenue management. 4G involves both capital and operating expenses, and those investments will have to be made simultaneously. CSPs will need to shift their perspective from cost to revenue management. For that, they will require more sophisticated policy and charging solutions.

The market dynamics of the Web2.0 will impact the LTE business models because it will be difficult to charge the user directly for the use of Web2.0 applications (that will run fast and smooth over LTE pipes) because Internet applications are associated with free usage. In view of this Telcos could exploit more indirect revenue sources : in a 'multi - sided' market structure the telco transactional platform can facilitate improved interactions and transactions between people and organisations : between advertisers and the end users.

Businesses can capitalise on 4G LTE for a wide set of applications, some of which are purely 'horizontal' while others are highly sector-specific, addressing needs unique to the industry. LTE's advantages are of greatest relevance to applications for personal communication and collaboration, CRM and project management. LTE will deliver improvements in the performance of many existing applications, and make feasible new applications that depend on reliable high speed or responsive data transfer.

Within the next decade, and probably by 2015, one trillion devices will be connected to the network – most not phones – moving the communications industry from quad-play or multi- play to "Tera-play". And LTE is one of the key enablers of a Tera-play world. For service providers, the benefits of Tera-play could be substantial with opportunities to drive revenue from an increasing number of high-value, multi-device, multi-service operators brand and product marketing requirements.

In a recent survey (Telecoms Intelligence BSS), 84% of operators revealed that they will invest in solutions for smart upsell offers, triggered by real-time context information, such as network usage, application access, location and more). This will help operators to maximize upsell opportunities and sales conversion rates as the offers are more relevant and timely.For example, a subscriber on a low speed data package, accessing Netflix, could trigger a high speed add-on offer; Another subscriber trying to access Facebook with no data package could trigger a data pass offer with unlimited access to Facebook. One operator successfully used this approach to stimulate data adoption by upselling real-time data passes (10MB for 1 day) to subscribers who had no data plan when they tried to consume any data. They used data passes to provide total cost control and remove any fear of bill shock .

The build out of today's 4G networks such as LTE requires a dramatic increase in computational resources to adequately deliver flexible telecommunications services to mobile subscribers. Yet business conditions also necessitate that new markets are approached incrementally. The challenge for telecom carriers is to reduce the cost of serving the first subscriber in small or cost-sensitive markets. The primary challenge in serving small LTE subscriber bases is that traditional core network architectures require high capital expenditures just to serve the first subscriber.

Networks, whether entry-level or full-scale, are traditionally built using separate network elements for each of several different functions. And most network elements have been deployed with a pair of carrier-grade servers to achieve redundancy with an active and a standby configuration. Thus, a new network with 10 network elements requires 20 servers just to provide service to the first subscriber. Furthermore, because the network is designed to eventually support a large population of subscribers, the servers would remain underutilized until the subscriber base grows to the expected population. The ROI for small and emerging markets has therefore been limited by these high capital outlays. High operating costs for maintaining the servers and providing data center floor space, power, and cooling have also hindered new service opportunities.

The Amsterdam Internet Exchange (AMS-IX) is the largest in terms of connected Autonomous System Numbers (ASN). The significance of an Internet Exchange is measured by the number of peering networks (Autonomous System Numbers) and and the Peak Internet traffic in Gigabit per second. AMS-IX is a mainport for Internet traffic more than Rotterdam and Schiphol are for containers and passengers respectively. London, Frankfurt, Paris and Amsterdam form the leading group of colocation data centres hot spots in Europe. Measured in colocation supply m2 per € bn GDP, Amsterdam exceeds all other cities. As a result Netherlands is hosting the top of the world's technology and Internet companies as gateway to Europe and the Internet such as : Facebook , Twitter , Netflix , Akamai , Amazon , Google and on and on. For this reason world's largest service providers and e-commerce companies have chosen Amsterdam as their #1 or #2 Internet Exchange position in the EU.

Large investments in data centres within the Netherlands by corporate multinationals like Google and IBM generate additional employment. Direct employment in the Digital Infrastructure sector adds up to 7,600 FTE, of which 90% in the hosting sector and 10% in capital intensive housing. Operational expenditures and investments in the housing and hosting sectors drive indirect and induced effects to create additional jobs. Combined effects for the Digital Infrastructure add up to 19,000 jobs in 2013 with a projected growth of 8% a year.The real value of the Digital Infrastructure sector, however, lies in its significant impact on the much larger Internet economy and broader digital society.

A continues interaction between Digital Infrastructure, service innovation and online usage drives growth in the online ecosystem. Digital Infrastructure is part of a much larger online ecosystem generating at least ~ €39 bn in revenue in the Dutch economy. Including private investments, government spending and trade, the Internet economy in Netherlands adds an estimated €34 bn to the GDP which is approximately 5.3% of the total GDP and steadily growing. There is a strong correlation between the Digital Infrastructure and e-commerce which shows that the former is a key enabler because E-commerce application are hosted in data centres and e-commerce traffic flows over the Internet exchange(s).

The employment generated by e-commerce in Netherlands is estimated between 100,000 and 140,000. SaaS and PaaS are two of the Digital Infrastructure's closest relatives, generating 5,700 jobs in the Dutch economy. Google has invested €600 million on a data centre located in Delfzijl, the Netherland.The estimated additional employment that the data centre will provide is 150 FTE from operations and a 1000 FTE at the peak of construction. The presence of most major global data centre providers in the Netherlands is prove of the country's attractiveness in terms of Internet Connectivity, availability of required electricity capacity , economic and political stability and highly-educated and multilingual workforce.

A large and complex ecosystem of companies and other entities compete, collaborate and cooperate to construct and maintain the interconnected network of networks that is the internet. The ecosystem works, and anyone can download a web page or video, or activate a mobile app, because of common standards and a shared understanding among participants of the benefits of a vibrant and growing economic system. Countries need energetic digital service sectors. They are drivers of social and economic development, job creators, talent magnets and the exports of the future. Robust digital service sectors depend on a complex ecosystem that includes adequate infrastructure and an investment-friendly business environment. The availability of mobile spectrum is one of the biggest, and most complex, infrastructure constraints especially in Africa. Governments must release additional spectrum – licensed and unlicensed – for private mobile use, as well as take steps to encourage spectral efficiency. New approaches to encourage harmonization and alternative deployment models are required.

At a high-level meeting during Mobile World Congress in Barcelona, senior leaders from eight major mobile operator groups, serving 551 million mobile connections across Africa and the Middle East, resolved to cooperate on network infrastructure sharing initiatives that recognise the profound impact of mobile broadband and Internet services on the citizens of both regions. The participating operators made this commitment in order to provide Internet and mobile broadband access to unserved rural communities and drive down the cost of mobile services for all sections of the population.

This initiative basically echoes the GSMA's call that telecom regulatory frameworks should encourage flexible commercial sharing arrangements and facilitate access to government-owned assets at preferential rates to help speed up the roll-out of new networks and support the business case to extend mobile networks into rural areas. Regulators should consider the competitive advantage that sharing of towers could provide in their respective markets. However, what they have to bear in mind is the fact that new and smaller operators will be incurring lease payments as an operating expense with relative lower risk, whilst the large and incumbent operators are still recovering the capital expense incurred in erecting the towers.

So what is really driving this network sharing phenomena apart from the altruistic motives of bridging the digital divide ?? Well how about the fact that increasing competition, along with investments in ever-changing technology, which has been pushing telecom operators towards new ways of maintaining margins. Since building and operating infrastructure is a significant cost for operators ,network sharing it is the ideal way to roll out infrastructure quickly and efficiently in low ARPU rural environments. operators can rely on a single set of infrastructure for their network. According to experts the estimated Capex savings resulting from tower sharing in the Middle East and Africa region amount to USD 10 billion. Quantifying and realising these savings requires a rigorous business plan and a meticulous execution controlled through appropriate contract governance structures and well-defined service level agreements.

Currently the most commonly shared infrastructure among operators is passive infrastructure, as it is easier to contract its set-up and maintenance. Sharing passive infrastructure only, means that newer operators still need to set up their own transceivers and other transmission equipment.Passive infrastructure sharing (commonly referred to as tower sharing) has attracted significant interest from both operators and tower companies. Companies like Helios Africa, American Towers and Eaton Telecom are already working to gain first-mover advantage by pursuing tower acquisitions in the region. Over the last 2 years, the tower business has grown into a fully-fledged industry in Africa and the Middle East.

Passive infrastructure sharing requires the consideration of many technical, practical and logistical factors although the principle is simple in theory. Any potential impact must be assessed and fully understood before sharing commences to ensure that there are no adverse effects on the operation of the site and the supporting network equipment and systems. Operators must consider items such as load bearing capacity of towers, azimuth angle of different service providers, tilt of the antenna, height of the antenna, before executing the agreement. Although, tower sharing enables new entrants to scale-up faster, it exposes established players to the risk of market share loss. Furthermore, the challenges of monitoring network performance and quality will increase as control over network roll out and equipment maintenance decreases.

As passive infrastructure business has evolved into a separate industry around the world, many tower companies in the telecom industry face several challenges. These include:

• High capital requirement: Tower deployment is a highly capital-intensive activity. The installation of each tower requires an investment of USD 55 000 to USD 75 000. Thus, tower companies the world over end up being highly leveraged

• Regulatory clearances: The first step should be to ensure that the regulatory authority is in favour of infrastructure sharing. Projects may stall because of delays in regulatory clearances. Apart from dealing with telecom regulators, tower companies also have to deal with other governmental bodies such as municipalities, forestry departments and environmental departments.Hurdles in obtaining clearance from a multitude of governmental bodies are often cited as reasons for delays in several site installations across developing nations. Since most of them are regional in nature, tower companies have to deal with quite a few governmental offices scattered across the country

• Operational cost optimisation: Although operational costs such as power and fuel are generally passed on to the operators, these are usually subject to agreed maximum limits. Thus, tower companies must work towards building controls to limit operational costs. Tower companies also face the problem of finalising the cost-sharing percentage and building a technology road map.

• Handling of local issues: Tower deployment and operation involves dealing with location-specific issues, including dealing with the landlord and local authorities, and running operations across a variety of geographies and terrains.

In the US , TowerStream has formed Hetnets Tower Corporation, which will offer wireless carriers and others a range of shared infrastructure services and access for mobile wireless Internet services. They believe the explosion in mobile data in urban markets is driving a migration to small cell architecture, and the major carriers are presently focused on the densification of their networks. With the rise of mobile data placing a tremendous demand on the networks of the carriers, TowerStream concluded that its Wi-Fi network can serve carriers' data offload needs. To serve this need, Hetnets Tower Corporation rents space on street level rooftops for the installation of customer-owned small cells, which includes Wi-Fi antennas, DAS, and metro and pico cells. Channels on TowerStream's Wi-Fi network are available for rent for the offloading of mobile data.

Concurrently there is a growing industry in green technology that specialises in producing energy from renewable sources or with zero or reduced carbon impact. Such technologies include solar power, wind power, wave power and bio fuels. Operators should be in a position to benefit from these technologies as the amount of power they can generate continues to improve. Vendors have already successfully trialled a combined solar and wind powered base station in various African countries, which not only reduces the environmental impact of the site but also makes it more feasible for operators to deploy sites in remote regions by negating the need for traditional power supplies or maintaining a fuel generator.

Network roaming can be considered a form of infrastructure sharing although traffic from one operator's subscriber is actually being carried and routed on another operator's network. However, there are no requirements for any common network elements for this type of sharing to occur. As long as a roaming agreement between the two operators exists then roaming can take place. For this reason operators may not classify roaming as a form of sharing as it does not require any shared investment in infrastructure. When roaming agreements come to an end they can be renegotiated either with the existing host network or another operator with minimal effort and transitional impact.

Network sharing is increasingly favoured by progressive policy makers as a way of ensuring more rapid provision of 3G services and on environmental grounds. On this basis the European Union has consistently ruled in favour of permitting passive network sharing and more recently also national roaming under the caveat that competition rules are respected. The sharing of sites and masts, national roaming and RAN sharing tend to impact coverage, quality of service and pricing of services to consumers positively, as the cost saving characteristics of infrastructure sharing allow for increased efficiency.

One of the largest commercial infra sharing deals occurred in August 2004 between Telstra and Hutchinson in Australia This was cleared by the ACCC (Regulator) who assessed the benefits outweighed the potential competitive impact. Telstra paid $450 million to Hutchison Telecommunications Ltd for a 50% share in ownership and operation of its 3G radio access network infrastructure. The cost to Telsra of building a network over four years would have been $900 million to $1.0 billion. Telstra stated the deal was undertaken to save on costs of entering the 3G market and that they scored a tried and tested network at half the cost.Surely the same logic would apply to 4G LTE networks in MEA.Ofcourse we have to wait and see in the MEA Telcos will actually realise the well known savings of passive sharing and whether some of these savings will be passed onto the rural consumers in form of lower data / voice prices. In spirit at least , infrastructure sharing is a step in the right direction for MEA telcos.

I bet you did not know that the Barcelona Mobile World Congress was the world's largest tradeshow certified to have a zero carbon footprint. The GSMA will certify the Congress as carbon neutral through the internationally recognised PAS 2060 standard. For several years, the GSMA has had climate change initiatives that have focused heavily on reducing waste in printed materials, encouraging the re-use and recycling of materials at the venue, utilising digital signage and electronic tools and working with Fira Barcelona, exhibitors and local partners to minimise the carbon footprint of the event.

To achieve carbon neutrality, the GSMA undertook activities to reduce the carbon footprint and will then purchase carbon credits to offset any remaining emissions. Carbon credits purchased by the GSMA will fund several CER (Certified Emission Reductions) projects certified by the United Nations Framework Convention on Climate Change (UNFCCC), including a hydropower project in China and a wind energy project in India, among others.

So what has this to do with us clever Telco people ?? Plenty.... unless you live under a rock you would know that telecommunications is supposed to reduce the need for transportation and the movement of people, as such total energy consumption should decrease in spite of the increased energy consumption needed for telecoms. Telco energy efficiency is such a critical issue that the GSMA Green Power for Mobile (GPM) programme, includes several initiatives such as awareness creation about the renewable technologies for telecom applications, CAPEX and OPEX analysis, vendor mapping and renewable energy market sizing.

The goal of GPM is to assist the mobile industry in adopting renewable energy sources, such as solar, wind, biomass, fuel cell or sustainable biofuels and hybrid power systems, in order to power an estimated 118,000 new or existing off-grid base station sites in the developing regions of the world. Reaching this target will reduce an estimated 2.5 billion litres of diesel consumption per annum and up to 6.8 million tonnes carbon emissions annually.

Due to an unreliable electrical power grid, Telcos / Tower Cos in Africa , India and other emerging markets use diesel generators, batteries and a variety of power management equipment to back-up the grid and ensure network availability. The growing cost of energy due to increasing diesel prices and concerns over rising greenhouse emissions have caused Telcos and TowerCos to focus on better power management methods. Mobile network operators (MNOs) spend approximately US$15 billion on their annual energy use. Therefore, it is no surprise that energy efficiency is becoming a strategic priority for them globally. As mobile use continues to grow, so does the demand for energy, particularly by the network infrastructure.So lets take a quick look at the REAL cost of Diesel.

A typical generator at a telecom site consumes 2.5 litres of diesel per run-hour. Add servicing ($0.76 per run-hour) and replacement ($1.05 per run-hour) costs to the fuel itself ($1.10 per litre) and you get a fully loaded cost of $4.56 per run-hour.In Nigeria there are currently about 24,500 operational base station sites:12,000 are connected to the grid, of which approximately 80% need generator backup for regular grid outages lasting anything up to six hours per day plus 7,000 are generator-powered 24/7.And of the remaining 5,500, the vast majority are diesel-battery hybrids, with just a handful of systems also using renewable energy.

For the purposes of our calculation, let's assume that on average they run their generators for 12 hours per day. Pull that all together and diesel-related costs for existing sites add up to a staggering US$485M per year. And when you consider the country needs to increase the number of base stations to 60,000 by 2018, it's putting increasing pressure on operators whose power-related operating costs are skyrocketing while their subscribers are simultaneously pushing for lower prices. And this is just for Nigeria – one country (all be it the most populated one) in a continent of 55.

Renewable Energy Technology (RET) solutions like solar photovoltaic, wind power, biomass and fuel cells are the technologies of choice for alternative solutions at telecom towers today. Hybrid solutions that combine diesel generators with RETs and batteries are being customised. Fuel cells are being installed as a standalone solution replacing the existing diesel generator. In a limited number of cases where electrical grid availability is close to 20 hours a day or more, the diesel generator at the tower site has been replaced completely by enhancing the existing battery capacity leading to improvement in economics and reduction of carbon emissions on site. Batteries are and will continue to be a key part of any backup power solution.

India is one country that has played a pioneering role in the field of energy efficiency. Some of the initiatives that have been implemented in India so far include passive infrastructure sharing, replacement of old base transceiver stations (BTS) with new generation BTS, usage of outdoor BTS, optimised cooling at shelter, usage of intelligent transceivers (TRXs), reduction of air conditioner load by using cold ambient air for shelter cooling and operating air conditioners using stored energy in the batteries to reduce diesel consumption and carbon emission are. In the last four years with the evolution of technology, the typical power consumption of BTS has dropped by about 60% .

Bharti Infratel claim their introduction of Free Cooling Units (FCU) used in place of air conditioners has contributed to reduction of 4.1 million litres of diesel usage annually after deployment across 6,318 of its 34,220 tower sites. Technologies like solar photovoltaic, wind power, fuel cell and other renewable or clean energy sources have been deployed in about 4,021 telecom sites in India. Approximately 1,000 Indus Towers sites use solar photovoltaic to augment the grid and diesel generated power.

Lets take another example : Nigeria and Ghana combined have a total of 30,000 + sites of which 50% are located in areas without commercial grid power , and mostly in rural and remote locations often with difficult accessibility for smooth and effective operations. Therefore , operation and maintenance of the network remains a big challenge affecting the cost of operations network availability and reliability of mobile telecommunication services. Network OPEX will depend on various operational factors including energy supply, equipment maintenance, operational efficiency and robustness. The right technology, robust systems, right supply chain integration coupled with regular monitoring and reporting will enable the MNO to achieve OPEX efficiency and improve profitability.

Operations and monitoring of deployed green power solutions is the most crucial part for guaranteed savings and expected performance.In addition the MNO's must embed robust operational practices and a monitoring framework in order to address the challenges and mitigate the risks of theft, vandalism, and ensure site security. Site security is a major issue as there have been several cases of damage to tower assets across the region. This risk has hindered MNOs from investing in green power alternatives for powering the network. Thefts of equipment and fuel pilferage have affected the OPEX of telecom sites.

Meanwhile among the most developed nations the NTT group in Japan has worked hard on energy conservation, prompting them to introduce new sources of energy in the telecommunication field as part of the "Save Power" campaign from 1987. A total of 282 solar power systems generating a total power of 4,745 kW and 18 wind power generators producing a total power of 781 kW were introduced. NTT Facilities, which has taken responsibility for energy system design, architectural design, and building and energy management, proposed the new concept of "Green integration." This concept involves global environment protection and is based on NTT Group's past experience and expertise.

NTT Facilities has Green data centers, are operated in an environmentally friendly way and construction of the facilities are based on the "Green integration" concept. NTT Facilities has developed a FMACS airconditioning system, by taking into account the indoor air current, that can effectively cool such equipment, and the system eliminates high temperature areas and reduces the amount of energy consumed for air conditioning. NTT also architected and deployed a MICRO GRID which combines various distributed power such as fuel cells, solar cells, and NaS batteries. The energy control system operates the distributed generators to control the influence on the commercial electric power lines wherein the micro grid is connected. This control system also optimizes the generation scheduling in terms reducing cost and environmental impact.

So what's the point of all this ?? GO GREEN OR GO HOME...unless you are fond of having 35% + of your monthly OPEX going up in diesel fumes !!

The release of additional spectrum is often used as a vehicle for introducing additional competition by the Regulators. From the perspective of spectrum regulators, careful spectrum management is required to ensure that sufficient spectrum is available to support not just the development of the commercial mobile market, but to support the continued operation of critical services such as Government, utility and Emergency Services that use radio spectrum on a daily basis. For the mobile operators spectrum is a critical resource , notably the ownership of lower band spectrum without which 4G will remain a myth in some emerging countries.

As we all know frequencies in the low band range 700MHz to 2.6GHz provide the optimal combination of propagation or coverage (the lower the frequency the better the coverage) and the ability to carry information or traffic (the higher the frequency the greater the data carrying capacity). Indeed some dense urban cities networks are now approaching the limit of network densification and additional spectrum and or new technologies may be the only route for alleviating network capacity constraints. Not to mention to stay in existence over the long haul !!

Unfortunately many mobile markets are no longer experiencing revenue growth as mobile broadband revenues simply offset declining voice revenues. To exacerbate the situation only smart Telcos have figured out how to monetise LTE.Telcos will need to consider the requirements of these air interface choices – such as the levels of handset/terminal take-up, as well as base station, antenna and transmission upgrades – when formulating their spectrum acquisition plans. In emerging countries the demand for high speed data services and delays in availability of Digital Dividend Spectrum has caused severe congestion on their current 3 G networks.

Understanding the value of spectrum to a business is essential for developing a spectrum strategy and participating in a spectrum auctions. The massive cash outlay for additional spectrum and the requirements to make a return on spectrum investments adds a layer of complexity to the evolving cost of data on HSPA and LTE networks. The substitutional nature of some spectrum bands requires a holistic approach to re-farmed 900 and 1800MHz spectrum strategy and valuation. The valuation process must consider stand-alone regional and / or block valuations and also packages of regions and / or blocks. When considering packages over stand-alone valuations the impact of scale must be included.

In respect of a spectrum auction an operator has to find an answer to three fundamental questions:How much spectrum do we need in different bands? The question relates to an assessment of spectrum need in the context of the growth in demand, notably mobile broadband. This needs to take account of the overall strategy, for example traffic offload through WiFi or Femto cells.How much is each block worth, i.e. what is the most we should bid for it? This relates to valuing each spectrum block in order to set the bid limit for the auction. This is quite separate from auction strategy. Clearly, if there is no bid limit, the auction will be simple because a bidder would simply pay whatever it takes to win the spectrum. However, such an approach may not result in the creation of shareholder value and may draw criticism from shareholders and the financial press and capital markets.

How do we obtain the spectrum as cheaply as possible? In any auction, the bid limits should be set before the start of auction. The role of bid strategy is to ensure the spectrum is obtained for less than the bid limit and at the lowest possible price. Depending on the auction format there may be an opportunity to influence the outcome and avoid negative effects such as aggregation risk (e.g. be stranded with unwanted blocks). This is addressed by examining the auction rules and developing a bid strategy which will be tested through simulations and mock auctions.

Operators may also have to consider mitigating strategies for a "no spectrum case" but if they face network congestion a range of mitigating strategies such as traffic shaping and fixed line off-load must be examined and incorporated into the valuation process.Operators must consider how regulation on net neutrality might impact their ability to shape traffic profiles and whether there are any long run cost implications of offloading to other players fixed networks.

Spectrum is often awarded through an auction process and recent auction designs favour a second price rule which means that bidders cannot influence the price they pay for the spectrum only the price that others pay. A bidder's valuation for a spectrum block is the price at which he walks away from a take-it-or-leave offer. Where aggregation risk is present valuations should be defined over packages, not just individual blocks. A valuation is conditional on information known at the time.

Depending on the auction format there may be dominant bid strategies or ways to avoid negative outcomes in cases where there is aggregation risk. This can be explored at theoretical level, through simulations and mock auctions. In theory a bidder enters the auction well prepared and the auction itself is a mechanical exercise. However, as the auction unfolds there will invariably be some learning which needs to be processed at the end of each day in order to be prepared for the next day's bidding.

The next big Spectrum " land grab " will take place in Africa (the perennial laggard in the broadband era) even as the Regulators dither and delay in the implementation of the Digital Switchover / Dividend. We predict that LTE will really come of age in Africa in 2015 by which time the 700 : 800 mhz and 2-6 Ghz spectrum becomes available thru auctions or beauty contests. At a joint ITU (International Telecoms Union) and ATU (African Telcom Union) meeting the outcome saw Africa become the first region in the world to be in a position in 2015 to cohesively and harmoniously allocate bandwidth freed up by the transition to digital television—the so-called 'digital dividend'— to mobile services in both the 700MHz and 800MHz bands.

However if Regulators in Africa dole out slivers of Spectrum to many " wannabe " Telcos in the false notion that this will decrease prices (as they did with Wimax) then expect the same mess : a host of under resourced " Pygmy " operators in each country setting up localised LTE networks with limited coverage and trying desperately to make a decent ROI. This scenario will do precious little to bridge the catastrophic Digital Divide in Africa. In a few years while the rest of the world will be on 5G African Telcos will still be boasting about their " 3 BTS me first 4G network " using their inadequate LTE spectrum allocations.

There is no doubt in my mind that developing countries Regulators will split the LTE spectrum into slivers over many bands. But here is the good nesws : Carrier Aggregation (in LTE A benefits operators with multiple spectrum positions, those with small pieces, and particularly operators that are combining acquired networks. The initial focus is on higher-speed services, but expect more deployments of 5+5MHz carrier aggregation as emerging markets deploy LTE in 2014.By combining blocks of spectrum known as component carriers (CC) , carrier aggregation enables the use of fragmented spectrum and allows LTE-A to meet its IMT-Advanced headline data rate of 1 Gbps. In simple terms bonding Spectrum channels together to create larger channels enables faster wireless services, and reduce opex and capex costs from running multiple networks. Believe it or not LTE A + CA is the 4 G technology for emerging markets.

A Spectrum bid requires a well honed strategy that factors in technical , commercial and financial parameters to balance a subtle equation that underlines 4G data networks. This is followed by bid strategy that will be implemented systematically along a project time line by a " tiger team " drawn from various disciplines. The acquisition of new spectrum and subsequent technology deployment results in massive Capex and Opex.So simply bidding without an all encompassing strategic plan and its flawless execution is a recipe for disaster....even if your Uncle is running the Regulatory Authority !!!

According to Ey ..market conditions for operators remain extremely challenging, with ongoing price deflation driven by competition from OTT providers and adjacent market players — such as cable companies and unbundlers — moving into mobile services. At the same time, fixed and mobile convergence is putting greater pressure on operators to widen their service propositions as cross-selling strategies gain more importance. Under these conditions, stressed operators see consolidation and transformational M&A as a route toward more rational market structures, which they believe can support higher levels of network investment. With global telecoms M&A deal volumes on an upward trajectory, our survey highlights the ongoing focus on these opportunities, with in-market deals to generate scale, widen product scope and reduce churn currently top of mind.

For operators, infrastructure migration road maps will also require careful consideration as a new range of technologies becomes available to support long-term migration to 4G, fiber and M2M environments. Meanwhile, the pace of M&A is likely to accelerate, with the current focus on in-market consolidation giving way to a new wave of cross-border deals as operators consider more ambitious routes to generating scale.

In one of the largest M&A deals this year was Telfonica's agreed to buy the E-Plus German wireless unit of Royal KPN NV (KPN) in a cash-and-stock deal valuing the unit at 8.1 billion euros ($10.7 billion) to become the country's biggest mobile-phone operator by customers. The Dutch phone company will get 5 billion euros in cash and a 17.6 percent stake in the combination of E-Plus and Telefonica Deutschland Holding AG (O2D), the Spanish carrier's German unit, which uses the O2 brand. As the EU commissioner in charge of the digital agenda, pushes reform in favor of a single European telecommunications market, carriers have become more emboldened to pursue deals. They are seeking to share expenses to build so-called fourth-generation networks to cope with rising demand for faster data connections

An analysis of the transactional rationale of the Telefonica deal provides valuable insights into the main elements for considerations of " best of breed " Telco M&A deal meaning covering the financial and non financial bases. Ofcourse there is no substitute for old-fashioned focus on the fundamentals of M&A: a clearly articulated and well thought-out strategic rationale for the acquisition becomes the yardstick by which to measure individual decisions that arise during the course of a transaction. Without one, decisions are made that end up being costly and inconsistent with the ultimate strategy chosen – or worse, require divestment of the entire acquisition years later as a 'bad deal'.

First Telefonica's desire to create a Digital Telco Titan : become a leading player with a combined customer base of 43m, 42% in postpaid and derive strong scale benefits with combined mobile revenue market share of 32% . O2 and E-Plus's combined customer base at the end of March would leapfrog Vodafone's 32.4 million and Deutsche Telekom's 37 million, according to data compiled by Bloomberg Industries. Germany has become the hottest battleground for telecommunications assets in Europe as demand for video and music delivered wirelessly and over the Internet increases, while voice revenue declines.

Second motivation was value crystallization through significant synergies. Telefonica is targeting NPV of synergies of €5.0–5.5bn, net of integration costs with projected Net savings from year 2 and Annual run-rate synergies of approx. €800 m; 75% of run-rate synergies by year 4. The deal will result in cost savings and revenue "synergies" of around 5 billion euros. Telfonica identified achievable synergies by rationalisation of distribution network ; increased efficiency in customer service costs leveraging best practices and scale ; better channel management and reduced overheads ; focused rollout on one common nationwide LTE network and improved quality from 3G network consolidation ; backbone, backhaul and core network consolidation, with reduced OpEx from network integration (rentals, power, maintenance, transport costs, overheads) ; site consolidation and rationalisation via reduction of around 14,000 sites ; increased efficiency by leveraging scalable transmission agreement with Deutsche Telekom and reduced SGA expenses by process rationalisation and a focus on become a more lean agile organisation.

Third motivation Telefonica wants a single LTE network to provide what they call the Best Mobile broadband experience. Key factors included giving customers to benefit from the best high speed mobile and fixed experience from a single LTE network and access to future-proof DT NGA network ; Tariff innovation, voice & video; mobile data bundling ; Strong multi-brand portfolio across segments ; Offering ICT / cloud solutions for business customers ; extensive distribution channel and outstanding customer service ;Leverage convergence through cross-selling / up-selling opportunities as well as profiting from digital innovation and scale from Telefónica's global capabilities (data centres , portfolio of OTT services and partnerships) .

Final motivation was value Creation for Telefónica Deutschland Shareholders . Here they were looking for enhanced financial flexibility (improving leverage) while maintaining an attractive shareholder remuneration ; maintaining conservative pro forma balance sheet with a projected EPS and FCF accretive from first year of full operation. In addition the M&A was all about investing in future growth while reinforcing geographical diversification, increasing exposure to an attractive market with a positive impact on Telefónica's cash flow generation profile.

Telfonica opted for the " Financing Without Increasing Leverage " motto meaning the deal is very positive for Telefonica from a business perspective while it doesn't affect its debt position. They have a Rights Issue in enlarged Telefónica Deutschland of €3.70bn. Telefónica subscribes prorate to its stake of 76.8%, €2.84bn + €1.30bn to KPN for 7.3% stake in the enlarged Telefónica Deutschland. Required total financing of €4.14bn is structured as 50-65% Hybrid, 100% equity under IFRS/ 50% equity for credit rating agencies and 20-30% Mandatory Convertible. Their objective is weighting around 2x incremental OIBDA, excluding synergies ; with Net debt/ratio preserved in short term for neutral to positive impact but keeping strong liquidity to maintain 24 months maturities for FCF generation till deal completion. Economic KPIs and cash flows must be consistent with real value creation. There is no place for speculation, particularly in these variable markets where sources of capital are skeptical, margins becoming tighter, and the consequences of missing forecasts are more direct.

As the folks at Ey rightly point out that a fanatical focus on due diligence of all aspects of the target's business and complete regulatory and market landscape is indispensable when there is so much money at stake . There is no substitute for old-fashioned focus on the fundamentals of M/A: a clearly articulated and well thought-out strategic rationale for the acquisition becomes the yardstick by which to measure individual decisions that arise during the course of a transaction. Without one, decisions are made that end up being costly and inconsistent with the ultimate strategy chosen – or worse, require divestment of the entire acquisition years later as a 'bad deal '.

Proactive Telcos clarify their M/A approach, organization and the way they source deals globally, and work closely with investor relations to ensure they have the right story to tell the capital markets—especially if they aggressively pursue new adjacent areas that have different value creation profiles or emerging market economies with majority low ARPU subscribers .Recent research by (JDSU / STL) has revealed an US$11Bn global opportunity for operators to monetize the data in their networks about places and people. The study concluded that demand for what it calls location insight services (LIS) will be driven predominantly by retailers that want to know more about local market trends and benchmark themselves against their competitors. Telcos are uniquely positioned to capitalise on LIS, as opposed to location-based services (LBS), which is proving more lucrative for over-the-top (OTT) service providers than telcos.

For some time the mobile industry has focused heavily on the opportunity presented by real time Location Based Services (LBS) for individual subscribers, a market that is estimated to reach $12.7 billion by 2014, according to Juniper Research. While there has been great success with LBS for apps targeted at consumers, many mobile operators have struggled to realize their share of this opportunity, with most of the revenue going to over-the-top (OTT) content players. OTT players lead the way in using real-time location data to provide location-centric services to consumers, such as special offers or vouchers.

By contrast, the Location Insight Services segment offers operators a new opportunity to monetize their location data. Telcos have a clear advantage over OTT players because they can aggregate huge volumes of anonymous location data over time and delivering value either directly to businesses, or via partners such as retail planners and advertising agencies. The underlying premise is that identification of repetitive patterns in location activity over time not only enables a much deeper understanding of the consumer in terms of behaviour and motivation, but also builds a clearer picture of the visitor profile of the location itself.

LIS plays to the strengths of operators because their engineers already collect anonymous location data for the purposes of analysing network performance and capacity planning.Various analysts have confirmed that there is a massive latent demand for location-centric information within the business community to enable the delivery of location-specific products and services that are context-relevant to the consumer. According to the Economist Business Unit, there is a consensus amongst marketers that location information is an important element in developing marketing strategy, even for those companies where data on customer and prospect location is not currently collected

LIS is an extension of existing software and analytics systems although data collected by these systems requires additional processing before it can be re-packaged into something marketable.This information can be shared with external systems and can be integrated with data warehouses using cost effective techniques. In many cases the intelligence can be directly used with business intelligence solutions.While commonly available cell level location enables some of the use cases, building level location intelligence from a carrier grade LIS system significantly increases the value. Examples of LIS include:

• Competitive Benchmarking (Retail) – previously unavailable intelligence on the profile of visitors to competitive stores
• Infrastructure Planning (Transport) – clear identification of "pinch points" on transport infrastructure and the precise times they occur
• Site Selection (Event planning) – evaluating previous attendee levels at a venue and attendance at competitive events with a similar audience profile
• Advertising Evaluation (Advertising/Retail) – determining the impact of advertising on store visits

For example , LIS platforms can enable mobile operators to share precise location data with transport infrastructure planners to help the understanding of where in the transport network heavy traffic occurs and when. This insight can be used to plan effective investment in infrastructure, and increase citizen satisfaction by improving transport network efficiency. LIS platforms provide the trend insight about which venues receive the best audience attendances given certain parameters, which can then be used to create a framework for predictive audience modelling. This enables event planners to more accurately assess the viability of venue locations, without needing to carry out time and resource intensive customer research.

Some Tier 1 Telcos have recognized the opportunity and publicly made noises about providing this insight. Last year Telefonica Digital unveiled a new division called Telefonica Dynamic Insights, which is tasked with monetising its vast data resources. Their first product, 'Smart Steps', will use fully anonymised and aggregated mobile network data to enable companies and public sector organisations to measure, compare, and understand what factors influence the number of people visiting a location at any time. These insights will help retailers tailor local offerings for existing stores and determine the best locations and most appropriate formats for new stores. A number of retailers are already helping with product development by providing user feedback. Smart Steps will also be able to help town councils measure how many more people visit their high street after the introduction of free car parking, farmers markets, or late night shopping.

Big data is one of the more fascinating developments in today's tech world: harnessing the huge wave of information that comes out of Internet-based networks and then trying to make sense of it. Mobile operators have huge repositories of data in their businesses : not just from people's activity on cellular networks, but from WiFi networks, too. LIS puts the power back in operators' hands allowing them to monetise the value of their unique asset, mass location intelligence, creating new revenue streams in times where traditional business models remain under extreme pressure. Hey guys, it is time to stop complaining about OTT marauders and take action to monetise one of your network's biggest yet most untapped asset : Location Location Location !!

---♠---

Telco Digital # 5 : The Business Agenda

The arduous task of creating fresh and viable business models in the hyper competitive Internet era bedevils Telco CxO's and their financial backers. All markets are becoming mature and emerging market growth will slow in the next few years. Have Telcos ever considered the nightmare scenario when Apple and Amazon offer handsets plus e SIMs. Contracts and billing could be handled via iTunes or Amazon accounts. The telco would thus become anonymous, a mere wholesaler of network capacity with no end-customer relationship of its own.

Most Telcos focus on matching and beating their rivals. As a result, their strategies tend to take on similar dimensions. What ensues is head-to-head competition based largely on incremental improvements in cost, quality, or both. So how do you leave rivals behind while sustaining spectacular growth for your company? Invent entirely new markets where no competitor has yet ventured. New business models are required where operators make money from new customers and ecosystem partners rather than exclusively from end-users.How can companies create breakthroughs in value and performance?

Companies that want to be successful in the current environment have to fundamentally scrutinize their business model on a regular basis and challenge its components if necessary. The overarching goal of a business model is to address a business opportunity in such a way that value is created for customers as well as for the company. A business model encompasses the addressed value potential, the customer interaction, as well as the value creation model.

Although many Telcos believe that they urgently need to build strong digital businesses, most are struggling to do so. Creating a Digital Telco means looking beyond traditional telco business models in the context of the changing telecom value network.The challenge for Telcos isn't that OTT companies outspend or " out imagine " them in digital innovation. It's that marginal cost analysis steers Telcos towards investments in capabilities that were relevant in the old basis of competition, rather than toward developing new capabilities relevant for the new basis of competition.

The mobile industry is undergoing a dramatic rethinking of business foundations and supporting technologies. In many ways, technologies such as cloud, software-defined networking and 5G result in a "software is eating the network" end game. This in turn will promote opportunities that are much larger than just selling voice and data access meaning digital commerce, advertising, energy services, smart home , e-health ,M2M , Connected Car , Big Data , IoT etc .It is for the Telcos to adapt , improvise , transform to profit from these opportunities. The Telco that is best able to connect the trinity of technology, content and segment will be able to reap superior profits.

Perhaps one of the most successful new age " experience " players to date is SK Planet, which was set up in 2011 by SK Telecom, Korea's largest wireless operator, to offer multiple add-on experiences for both retail and business subscribers. They include MelOn, already Korea's largest music portal, with 17 million subscribers, has also been launched in Indonesia; 11st provides an e-commerce platform with related advertising and marketing intelligence services.It is now the country's second-largest e-commerce platform and largest player in mobile commerce; "T ad" is a mobile ad platform that enables personalized ads on mobile apps running on smartphones and tablets; "T map," a GPS-based navigation service platform with more than 10 million subscribers, also offers location-based services to businesses.

Operators need to realize that extending connectivity alone cannot keep them afloat. Instead they require software, device and service strategies that can add value and at the same time differentiate them from competition.In the future, the primary operational mode for large players might well be as aggregators of massive services. The key is to open the platform and gain as much partner power as possible. This is the fundamental reason why concepts like Web2.0 and P4P became important.

NTT DoCoMo pursued the concept of Value innovation : a new way of thinking about and executing strategy that results in the creation of a blue ocean and a break from the competition. More importantly, value innovation defies one of the most commonly accepted dogmas of competition-based strategy — the value-cost tradeoff.In order to achieve its transformation strategy, KPN Netherlands mapped out four concrete objectives: fixed and mobile service convergence; full utilization of current network resources; considerable decline in CAPEX and OPEX; and delivering a variety of new services, which are built on an All-IP network.

Through creative partnering and innovative risk sharing options, new managed services and outsourcing business model options provide the framework for creating a next generation enabled portfolio of services for consumers and enterprises ready for 2 sided business models. Instead of short-term tactical advantages, the focus is firmly on long-term strategic gains by identifying blue ocean market segments where competition becomes irrelevant . And only then will Telcos be able to reap the full benefits of managed services via trusted partnerships.

According to VisionMobile , to grow Telcos could either...

- diversify, which means branch out into new markets by investing their profits into that new market, and try to turn that new market into a source of profits in its own rights; or...
- grow asymmetrically, which means branching out into a new market with the intention of not turning huge profits within that new market, but rather to drive profit in the core market.

Asymmetric business models cross industries and force profits to migrate from one market to another.Google for example uses them to disrupt industry after industry: from mobile (Android), to television (Google TV and Chromecast), enterprise software (Google Apps), personal computers (Chromebook), travel (Google Flights), energy (Nest) and transportation (Android Auto and self-driving cars).

The new generation of messaging apps have asymmetric business models, they can sustain free services indefinitely and forever change the dynamics of the mobile messaging market. WeChat, Viber, Line and others monetise by using their platform for e-commerce, selling digital goods (stickers, games), physical goods and services (like taxi rides). They don't have to charge for the messages or even voice calls to be commercially successful.Telcos need to embark on a series of such changes in order to ensure that they can build upon their successes in delivering telecoms services. In the end, they will need to ensure that they continuously challenge established models and notions on their role if they are to truly innovate their business model.

For some time now the OTT's have been viewed as the anti-carrier Antichrist. Just recently the CEO of one South Africa's leading MNO's told a tech news agency that over-the-top (OTT) services like WhatsApp and Skype were unfairly benefiting from his company's costly infrastructure. He warned that his network was not prepared to spend billions on its network just for the OTTs to have a "free ride". Another Telco was quick to declare that OTT services were "skimming" its voice revenues. Without these extra revenues, the company claimed, it cannot afford to provide telecommunications to poor rural areas, where it runs at a loss....boohoo...Did these guys just wake up because the OTT invasion began 5 years ago ??

It all seems so unfair that OTT players make their money by loading more and more traffic on the operators' networks, thus causing the operator to pay for network capacity upgrades to facilitate the OTTs' increasing VoIP and IM traffic that was the root cause for decreases in the operators' core voice and SMS revenues. But if truth be told circuit switched voice and SMS are on the way out, and as operators move towards becoming fully IP based digital services providers. So some clever Telcos are entering into collaborative agreements with OTTs in order to generate new revenues and provide a deeper range of services to their customers.... instead of making dire predictions and pathetic threats !!

While working with OTTs is still at an early stage for many operators the $$ opportunity is to not view OTTs as a threat but to use them to build more comprehensive offers that customers want and help to build loyalty. Instant messaging is on the rise and texting is starting to see its first decreases in usage since it was launched. In August 2014 research firm Deloitte reported that the average person sends seven text messages a day, compared to 46 instant messages. SMS texting is forecast to fall from 145 billion to 140 billion by end of 2014. According to the survey of UK consumers almost a quarter of smartphone owners use five or more messaging apps.

There are many operators who have started selling application service passes which provide a low cost means of using an OTT service from a smartphone. For example, in Kuwait, Ooredoo offers a WhatsApp service for only 750fls (2 euro) a month. This is a good example of offering a low cost service limited only to a specific application to generate a new income stream and encourage mobile data adoption. Ooredoo promote this service as an alternative to trying to find a free Wi-Fi zone. As well as offering low price services for one app, operators are also zero rating popular services, such as Facebook or Twitter in order to get customers using their services.

In June 2014 TMF ran a survey of 67 operators looking at uses of real-time convergent charging and policy management. In this survey 56% of operators said they are offering zero rated deals, under which the use of services such as Facebook and WhatsApp don't count against customers' data allowances.

Operators can look to work with OTTs in order to develop new revenue streams to replace the traditional texting and voice revenues. It's not a case of operators 'versus' OTTs. Operators have some key assets that can make them very attractive to OTTs as a route to market and develop win-win partnerships. These include but not limited to :

• Established customer relationship often built up over years

• Established financial relationship – regular billing and subscription models, and prepayment for services

• Flexible charging, pricing and billing options – ability to provide innovative pricing plans which can help add value to the OTT's services

• Ability to prioritize the delivery of certain services – e.g. in a congested network operators can apply traffic prioritization

• Ability to differentiate service delivery – e.g. tiered service delivery: supply certain video traffic over LTE and offload others to slower Wi-Fi networks to avoid network congestion

• Use big data and customer analytics (e.g. cell site location, customer behavior) to upsell more relevant services and finely tuned advertising

E Plus in Germany and China Unicom have launched prepaid SIM cards that lead with OTT services. The WhatsApp SIM card from E Plus and the WeChat SIM card from China Unicom offer mobile voice minutes, a data tariff and zero rated OTT IM services. Looking at the E Plus example , which was launched in April 2014, this offers a prepaid card which costs €10 / month and the main message is WhatsApp is always free, even when the subscriber runs out of credit (in the 30 day period). This is a major benefit over other networks who charge for the data required for WhatsApp (which is installed on 90% of all smartphones in Germany, giving a potential user base of more than 30 million people). By working with this OTT E Plus can leverage the popularity of WhatsApp to attract new customers and open new revenue streams.

China Unicom has worked with the OTT IM service WeChat to offer a WeChat China Unicom SIM card that offers additional services over the traditional WeChat mobile IM service. These additional services include increased group chat limits and extra free icons. By differentiating the WeChat service available with the China Unicom WeChat SIM there is another clear incentive for WeChat mobile customers to switch to China Unicom. This service was launched in August 2013 in Guangdong and it was reported that in one month it gained 1 million customers.

The most recognizable players in streaming music , Spotify, Beats Music, Deezer and Napster have all partnered with mobile operators. his may help explain Spotify's increase in paying customers from 6M in March 2013 to 10m in May 2014. Deezer has probably been the second most prolific in terms of mobile partnerships and they say their paying subscribers increased by 4M to 5M between May and November in 2013. Sprint have added Spotify to their Framily (friends and family) plan. All Family customers get a six-month trial of Spotify and they will get Spotify at the discounted rate of $7.99 a month. For family groups of 6 to 10 members, the rate falls to $4.99 a month.

Telecoms operators have a choice - they can collaborate or compete with OTT service providers. Operators in the Middle East are taking both approaches. While incumbent operators in Saudia Arabia (STC), the United Arab Emirates (Etisalat) and Qatar (Ooredoo) are competing with OTT providers by either extending their IP TV services to OTT platform or by emulating popular OTT services, newer entrants to the market such as Saudi Arabia's Mobily and Nawras in Oman have collaborated with WhatsApp to launch specific packages. Because launching in-house OTT services or partnering with OTT players can be a double-edged sword, taking the precautionary step of adapting the services to the operator within the competitive context is highly recommended.

The recent interconnect deals with AT&T; Verizon and Comcast that Netflix has struck illustrates that in order to deliver video at the speed and quality that the content providers (and consumers) want then telecoms operators and customers cannot be expected to pick up the delivery bill all the time. Perhaps this illustrates that content providers are recognizing (and will pay for) differentiated service delivery. With over 50 million subscribers Netflix is enjoying substantial growth and has established payment of a monthly subscription fee. With viewing behavior changing and more consumers watching videos and TV shows on tablets and smartphones, companies like Netflix are seeing the potential for working with mobile operators (and vice versa). If operators are looking at partnerships with video / TV content providers then the first thing they need is real-time data collection, rating and notification systems.

Rather than just have all video traffic on a LTE network an operator could offer options to offload onto available Wi-Fi at a lower (or no) cost. This is done by having on device ANDSF (Access Network and Discovery Function) software which enables operator controlled offload by assisting devices to discover access networks in their vicinity (e.g. Wi-Fi) and provide rules to prioritize and manage connection to all networks. This allows operators to dynamically control and define preferences – that is how, where, when and for what purpose a device can use a certain radio access technology – e.g. under what conditions is traffic offloaded to Wi-Fi.

Having data delivery charges picked up, or subsidized, by the content provider or by advertisers could be a possible way to take the sting out of data heavy content charges. The delivery of media content has been traditionally subsidized (television, cinema, newspapers, radio) and it could be argued that this will not change for digital content delivered on mobile networks. If a customer was to watch a 30 minute TV show each day / month on a LTE network then the data charges would be around US$70 / month (for 4GB). Most customers would eschew this. If part of the cost was picked up by advertisers then maybe the retail costs could come down, with the resultant increase in active subscribers, content usage, and advertising revenues.

However the systems required to support much more diverse and dynamic business will need to be very different from the traditional legacy BSS that were designed to collect, rate & bill for circuit switched voice calls and SMS text messages in batch mode. Legacy BSS systems are the perfect recipe for bill shock and some very angry customers besides being inflexible when it come to dynamic service creation , delivery and monetisation.

Many telecoms companies are now pursuing multi-play strategies in which they offer a number of services, including mobile and fixed voice services, broadband and pay-TV. Vodafone has been active in this market with several high profile acquisitions in Europe, while the deal between BT and EE in the UK will also create a quadplay powerhouse. CCS Insight research suggests that consumers are interested in signing up to companies offering this range of communication and media services if their offering is good value. The appetite for multiplay services may also be driven by people owning an increasing number of devices.

Telcos today are facing margin pressures through more intense competition, ARPU erosion, customer churn and cost issues. While designing new business models, Telcos can leverage their network capabilities, such as mobility, messaging, location, presence, profile and call control, and combine these with internet-style services such as social networking, search, advertising, direct marketing and mapping, thereby enabling richer, more compelling and more personalised services than the Internet players can offer.

Furthermore, by exposing these capabilities in a secure, controlled and automated manner, Telcos can generate revenues from selling service enablers, as well as their own services, allowing them to fully exploit their network assets. In light of the aforementioned Telco Execs need to understand that :

1. The design and bundling of applications, content and devices to generate revenue from broadband networks is based upon a deeper understanding of the customer's data consumption habits

2. The business and technical logic underlying services delivery platforms because telecoms networks have evolved from voice-centric "legacy" technologies such as SS7 and IN towards data and multimedia-centric technologies based on IP, such SIP , Daimeter and IMS

3. The critical role of converged billing and CRM engines and how to convert BSS/OSS into revenue generating assets and the need to introduce attractive, profitable new services to subscribers with minimum time-to-revenue while controlling costs

Telcos can greatly benefit from implementing convergent customer care and billing systems because investing in a new stovepipe billing system for each type of service is an expensive and obviously sub-optimal proposition. The systems would help them bring new services to the market quickly, enabling them to improve customer loyalty and reduce customer churn, especially in an environment, where customers jump from provider to provider to get the best deals.

Big data has been a headline theme in the technology and mobile space for some time.Telcos all over the globe are seeing an unprecedented rise in volume, variety and velocity of information ("big data") due to next generation mobile network rollouts, increased use of smart phones and rise of social media. Telco operators have historically focused on managing the network with little visibility to the impact it has on the customer's experience. Which means the operator was forced to work with snapshots of network data in fragmented views or at a summary level in order to plan network capacity or provide information to customer care and marketing about customer transactions until now !!

Big Data technologies, and in particular their analytics abilities, offer a multitude of benefits to telecom companies including improved subscriber experience, building and maintaining smarter networks, reducing churn, and generation of new revenue streams. Mind commerce, expects the Big Data driven telecom analytics market to grow at a CAGR of nearly 50% between 2014 and 2019. By the end of 2019, the market will eventually account for $5.4 Billion in annual revenue.

Mobile commerce is one particular area where operators and service providers can potentially deliver tangible benefits from the application of big data analytics. The growth of m-commerce is creating large amounts of information on consumer behaviour and choices, which can be used to offer more personalised services and offers. SK Planet (Division of SK Telecom) have stated that "our Cash Bag m-commerce portal should generate $9.3 billion in revenues this year, and by using big data analysis we can provide customers with a much improved experience, and not based simply on offering the lowest price."

Big data analytics solutions enable service providers to analyze real-time location data over time for opt-in subscribers to understand subscribe lifestyle. Combining lifestyle and mobile profiles with subscriber usage and digital behavior allows service provider to create targeted offers for opt-in subscribers. With a majority of subscribers using smart phones to access data services as well as voice, mobile network operators are seeing explosive growth in traffic levels across their networks. In addition, the mobile network operator environment is fiercely competitive, with the ability to attract, retain and grow valuable subscribers being key. Increasingly, the provision of high quality customer care is an important component in the marketing mix and in retaining subscribers.

The growth of connected devices, particularly in areas such as the home or in the car, presents new opportunities but also challenges for operators and other ecosystem players. Users may be willing to share data with service providers but on the basis that the data is used securely. This year the GSM industry introduced a standardised mobile identity solution that aims to become the de facto single sign-on tool that consumers could rely on to authenticate themselves in both online and offline environments. This initiative is set to stimulate adoption of mobile services that rely on absolute confidentiality, such as healthcare, government and banking.

The Gurus at Strategy & believe that many types of data are potentially available to operators — and certain sets of data might be combined to open up new business opportunities in areas such as campaign marketing and fraud prevention. Operators could generate more accurate and personalized offer recommendations for existing individual subscribers by combining internal structured data, such as how and where each subscriber uses his or her phone, with external unstructured or semi-structured data from social media platforms (for example, Facebook and Twitter).

By correlating internal location, usage, and account data with external sources such as credit reports, operators could significantly increase the detection of fraudulent activity such as looping or call forwarding on hacked PBXs (private branch exchanges), or fraud involving the swapping of SIM cards, and improve the overall accuracy and efficiency of their efforts to recognize patterns of fraudulent behavior. Imagine having the best of both worlds ? Having the tools to analyze the growing amount of data and content your business is generating, and also finding ways to make it profitable. If you are astute then this deluge of data isn't a threat; it's a serious opportunity to take your telecom business in a new, exciting, and yes, profitable direction !! Here comes Enterprise 2.0...

In its most basic form, Enterprise 2.0 is about communication. When information is free, people can get more feedback and input (collaborate), react more quickly (agility), and make better decisions. This is the opportunity inherent in Enterprise 2.0: a more efficient, productive and intelligent workforce.The current crop of Collaborative solutions focus around unified communications (instant messaging, web conferencing and VoIP for example), working in teams, sharing documentation and knowledge, working with (self-service) portals and working with social collaboration tools.

Organizations are beginning to take advantage of social collaboration aspects like communities, blogging and wikis to connect with external parties like partners, customers and local government. A survey done by McKinsey & Company showed that companies that benefit most from B2C/B2B collaboration are:

- Networked organizations;
- Business to business organizations;
- Big companies (> $1 billion revenue);
- International companies;
- Decentralized organizations.

According to Industry experts there are three fundamental ingredients to be successful with E20/Social Business (or any major corporate initiative): Adequate resources/budget, organizational commitment, and a business problem to solve. Missing any of these greatly slows down and/or blunts the outcome of the effort. The top challenge is culture change. You can drop social technology into any organization, but you can't suddenly expect that employees will adopt the way that social media works or that business processes or traditions will automatically change.

Social is a new way of operating (observable work, openly participative processes, co-creation) and this requires conscious effort to change our thinking and the way we function. Other top challenges include enterprise apps with overlapping features (e-mail, CMS/DMS, IM, unified communication, enterprise microblogs, customer forums, CRM, etc.), underinvestment in community management, and lack of executive understanding or buy-in.Web 2.0 is the term for web-based tools and services that allow for – and even improve with – user participation. The most well-known examples of this technology are found in sites like YouTube , Facebook , Wikipedia and Amazon, where users to find and connect to what they are looking. Social media tools like blogs and microblogs (Twitter) opened up the world of media and publishing to anyone with an internet connection – or a smartphone.

Social network tools help staff find the right individual or group of people.Tagging and rating provide a straightforward way to find content and make judgments about what to look at. Blogs and wikis are natural collaboration and communication platforms.Giving employees the freedom to speak their mind and voice ideas is required for there to be a harnessing of collective intelligence.One of the biggest car lease firms worldwide had a clear vision on the use of social collaboration, both internally and B2C/B2B. In this vision it outlines the many strategies it intends to use for social collaboration.

* Launching an employee community
* Engaging in social recruitment
* Social software enabled car remarketing
* Launching fleet management communities
* Social software enabled car quotations
* Launching driver and supplier community
* Launching a supplier community
* Conducting online reputation management

A Global management consulting, technology services and outsourcing company implemented off the shelf social media platforms technology to introduce knowledge sharing communities and social networks. Blogs and wikis function as collaboration tools, and as such, they have uses mainly in sharing "unstructured" information associated with ad hoc or ongoing projects and processes, but not for "structured informational" retrieval.

However, Shell has started converting its official documentation to wikis, because this enables that company to make documentation updates available in real time and allows non-editors to contribute to the documentation. In this process Shell restructures the paper documents to a set of on-line wiki pages. Their key challenge was to get key stakeholders aware of how social computing can solve business problems and be integrated into business processes. The business case was based on the following metrics:

- Finding people and identifying experts;
- Finding information;
- Reducing the need for travel;
- Speed up the decision making process.

Current Telco mainstream offerings to the Enterprise market are based around capacity and hosted services, sometimes complemented by IT outsourcing projects. Mass-market consumers and Enterprise customers alike are increasingly demanding rich, portable, personalised, access and device-independent services from their Telco Service Providers.Telco 2.0 and Web 2.0 components creates more value to the Enterprise . For example Telco resources can be embedded with the Enterprise applications to identify the real time location and distribution of a service engineer's customers (using Google Maps and Location feeds) to view the geography of the area covered.

In the UK, BT (British Telecom) has become one of the country's strongest proponents of enterprise 2.0. The company has introduced a raft of social media tools, including a huge Wikipedia-style database called BTpedia, a central blogging tool, a podcasting tool, project collaboration software and enterprise social networking. With SDP / IMS platforms Telco customers can become part of the social networking phenomena by complementing these content based sites with telecom capabilities such as anonymous calling (whisper calls) and real time updates showing the physical location of friends and contacts within the community group. Whilst Telecom Web Services standard will revolutionise Telco service offerings to both the Large Enterprise and the SMB market, it is important not to overlook the benefits of being able to offer fully hosted services.

These include "Virtual PBXs" and "Virtual Contact Centres", plus a suite of complementary services to customer's own installed platforms such as "Mobile PBX Extensions", "Multi-line" services for handsets and of course "Voice Call Continuity" to provide Enterprise roaming in WiFi hotspots.The availability of Enterprise 2.0 tool combined with high speed networks smartphones and cloud computing will unlock fresh new revenue streams for agile Telcos and CSP's in the Enterprise / SMB markets.

We are on the cusp of a management revolution that is likely to be as profound and unsettling as the one that gave birth to the modern industrial age. Driven by the emergence of powerful new collaborative technologies, this transformation will radically reshape the nature of work, the boundaries of the enterprise, and the responsibilites of business leaders. Essentially, the future of the convergent industry is in service provision with globalization and personalization spurring consumer demand. The upsurge of convergence has brought not only opportunities, but unparalleled resource advantages to operators. However, whether an operator can go with the flow and thrive in the new environment depends not only on reformulating strategy but also on its timely and effective implementation.

Ever since strategic management was introduced to economy field, its main purpose is pursuing competitive advantage in the existence market. Blue Ocean strategy is a totally new strategy compare with old ones.In their landmark book strategy gurus Kim and Mauborgne (2005) divide the market universe into two parts: red oceans and blue oceans. "Red oceans " is described as all the industries in existence today which is the known market space. " Blue oceans " refers to all the industries not in existence today which is the unknown market space. In red oceans, the market boundaries is clearly identified, the market space gets crowded, prospect for profits and growth are limited, competition in red oceans turns to be bloody. In contrast, blue oceans are defined as a new space in which market boundaries and industry structure are not given and can be reconstructed. In blue oceans, competition is irrelevant since the rules of the game haven 't been set. Very few incumbent telecom providers has put into place any Blue Ocean Strategies. Yet Blue Ocean Strategies have made the Circus, Wine, Gaming, Airline, etc. industries exciting again, so why not apply it to the telecom market ?

Blue ocean strategy is about creating and capturing uncontested market space, thereby making the competition irrelevant. For example, NTT DoCoMo was the first company to make money out of the mobile internet. In a very competitive industry engaged in a technology race and strong price erosion, NTT DoCoMo was able to achieve superior performance when it launched its novel i-mode services in February 1999. It was an immediate and explosive success in Japan. As with NTT DoCoMo, the goal for a firm's blue ocean strategic move is the pursuit of value innovation — a leap in value for buyers and company alike. This comes from simultaneous pursuit of differentiation and low cost. NTT DoCoMo pursued the concept of Value innovation : a new way of thinking about and executing strategy that results in the creation of a blue ocean and a break from the competition. More importantly, value innovation defies one of the most commonly accepted dogmas of competition-based strategy — the value-cost tradeoff.

In order to achieve its blue ocean strategy, KPN (the largest Operator in the Netherlands) has mapped out four concrete objectives: fixed and mobile service convergence; full utilization of current network resources; considerable decline in CAPEX and OPEX; and delivering a variety of new services, which are built on an All-IP network. These services would include: multimedia personal communication services such as, voice, video, photo, data, message, and PTT; IP corporate communication such as, an IP private line, IP PBX, and multimedia service; and multimedia entertainment such as, games, IPTV, the worldwide Web, and Portal.

Operators need to realize that extending connectivity alone cannot keep them afloat. Instead they require software, device and service strategies that can add value and at the same time differentiate them from competition.In the future, the primary operational mode for large players might well be as aggregators of massive services. The key is to open the platform and gain as much partner power as possible. This is the fundamental reason why concepts like Web2.0 and P4P become important.

Many companies, particularly technology firms, do tend to continuously add small features to their products in an attempt to differentiate themselves from the competition through a continual process of "incremental innovation." Mobile telephone companies are particularly guilty of this, yet each additional "design feature" detracts value from the buyer as the phone becomes increasingly difficult to use. A study from Eindhoven University found that in the US nearly half of products returned by customers for refunds were in perfect working order, their owners just couldn't figure out how to use them.Innovation without value tends to be technology-driven, market pioneering or futuristic design that may shoot beyond what buyers are ready to accept and pay for. Value without innovation tends to focus on value creation on an incremental scale that, at best, improves value but is not sufficient to stand out in the market.

For Telcos , most value customers are real-name customers, unlike Google's anonymous and even nameless customers. Only operators can obtain data on user behavioral patterns, which is a huge advantage over other competitors.Hence, operators are in a position to help their partners promote their products and recommend suitable products to the right customers. They can match massive services to millions of users. This is the service aggregators core competitiveness, yet most operators have been asleep at the wheel and have not collected, collated or utilized the data.

Many applications, such as home security protection, health and medical care, do not merely involve information distribution and interaction, but require hardware deployment and maintenance as well. The OTT providers find it hard to provide delivery or services in a large area and operators with massive service teams can help them and profit as well. The OTT providers lack branding power and a strong credit rating. When users purchase their services, a guarantee is needed.

Operators with long-term operations experience and a good credit and credibility rating can play the role of guarantor. This means that operators must qualify suppliers and control risk. Similarly, when users pay for a service, they usually will not trust an unknown service provider. A third-party, reputable platform is needed for completing settlement and payment actions. This is where operators' existing mature billing platform can really shine.

Salesforce.com's strategic moves provide an exemplary demonstration of how a company can effectively create and renew its blue ocean in the B2B field by value innovating its single business on the product, service and delivery platforms alternately. By 'de-segmenting' the market and looking at exceptional buyer value across segments and looking for commonalities across non-customers, Salesforce created a new mass of buyers that traversed the traditional segment boundaries. Had they wished to make incremental changes in the industry they would probably have offered a traditional client server model with some element of web-based access as well.

Salesforce however did not go for that route. Instead, they decided to launch a new concept around the mantras of 'success not software' and 'low cost, good enough', which completely reinvented the way the industry thought about CRM and challenged the traditional value-cost trade-off that buyers were typically used to. In keeping the first principle of Blue Ocean Strategy, Salesforce broke out of the conventional wisdom trap and pioneered to create a new value proposition that forced the market to wake up and listen.

Pureplay broadband access – the mainstay of traditional telcos' business today – will remain the cornerstone of digital communication in the future for both landlines and mobile communication.But Telcos should build an open digitised platform, utilize their long-term operational resources and experience to integrate more and better services and provide desirable services to end users. By doing so, operators can gradually move their commanding positions from network connections to their own exclusive services or the services with the best user experience.The "right" strategic orientation for each telco depends on five key levers:

1) Personalization of service ecosystem and the customer experience
2) Uncompromising defense of relationships with end customers
3) Cost-efficient broadband network build-ups
4) Realignment and radical streamlining of operating models
5) Financial resources to drive digital transformation and consolidation.

If Telcos do their homework with respect to broadband access strategy, customer experience , transform themselves into a cost efficient lean operating model backboned on Omni-digital,they have a real fighting chance to recover the ground lost to the OTT players ...instead of complaining they have been relegated to " dumb pipes ".YES ..you are dumb if you don't learn how to swim in the Blue Ocean.Whatever the future holds, all Telcos are called on to refocus and realign – and to do so fast !!

Business model innovation is the key to unlocking transformational growth--but few executives know how to apply it to their businesses. In "Seizing the White Space," Mark Johnson gives them the playbook. Tackling the white space challenge often requires a dual focus exploring the intersection of multiple emerging trends and defining how the convergence of two or more technologies or capabilities might satisfy powerful latent consumer and/or customer needs. Whatever the opportunity, you must continually engage in rapid learning and decision-making around new consumers, customers, partners, suppliers, competitors, business models and other emerging marketplace dynamics if you are to achieve success.

Like any blank canvas white space can provoke fear and hesitation. Many companies are reluctant to enter any white space because of the unknowns. White space can cannibalize existing products or services, it can require extensive system design and support, and in some cases it can require very different business models.

As markets mature, competition intensifies; new technologies are invented, and new consumer behaviors constantly emerge. Organizations need to actively look for new sources of differentiation; white space mapping is becoming an important strategic exercise for organizational learning and strategic planning. For a company to remain relevant over the long term, it must respond to these shifting conditions intelligently and white space mapping needs to be part of its strategic planning efforts.

Amazon's business model innovation certainly allows it to deliver a diverse portfolio of customer value propositions that serves as its main competitive advantage. Culturally, a continuous focus on business model innovation keeps the company connected to its entrepreneurial roots — an advantage that should be coveted by even the largest of companies.At the end of the day, "It's all about customers." As Amazon demonstrates, even when customers have many choices, with business model innovation, it is possible for revenue and growth opportunities to flow from the basic way a business is put together — even without the use of drones.

The Telco 2.0 Initiative Gurus believe that the emergence of the New Mobile Web is creating a white space for operators seeking to (re)build their role in the digital marketplace. The New Mobile Web is a term used to describe the transformed mobile Web experience achieved through advances in technology; HTLM5, faster, cheaper (4G) connectivity, better mobile devices. The New Mobile Web will lead to a shift away from native (Apple & Android) app ecosystems to browser-based consumption of media and services. Web RTC is a technology that promises to bring voice and video into the web browser. At the same time, it offers an excellent opportunity for Telcos to innovate by extending telecom services into the open web and freeing voice from closed telecom networks.

"Software is eating the world," says Marc Andreessen, co-founder of Netscape, who himself was once on the receiving end of competition from Microsoft. For most industries it's not a matter of 'if', but a matter of 'when' you will face a software-driven competitor with a unique business model.Low (or zero) marginal cost of many modern technology products is the key enabler for the success of new age business models. This is why these business models (pioneered by the software goliaths)are different from traditional loss leader and bundling tactics. As these business models are self-sustained, the new age competitors fundamentally alter the market dynamics in the "victimized" markets. The new market conditions are often rendering the business models of incumbents unsustainable.

Lured by the promise of attracting higher-ARPU smartphone users, Telcos have worked hard to flood the market with smartphones at a wide range of price points. This strategy served the short-term goal of boosting the connectivity business, but may have jeopardized the long-term competitiveness of the service business by surrendering the customer ownership associated with authentication, user identity management and billing services. As the basis of competition changes to "choice and flexibility", vertical integration (ie : notion of all-in-one telco spanning network operations, telephony, messaging, data access, user identity management, authentication and billing, as well as distribution and retail) has lost its competitive lustre. The lack of flexibility inherent to vertical integration explains why Telcos lost out to smartphone and Internet platforms in the areas of location services, authentication, single sign on, user identity, and billing.

Although many Telcos believe that they urgently need to build strong digital businesses, most are struggling to do so. Creating a Digital Telco means looking beyond traditional telco business models in the context of the changing telecom value network.The challenge for Telcos isn't that OTT companies outspend or " out imagine " them in digital innovation. It's that marginal cost analysis steers Telcos towards investments in capabilities that were relevant in the old basis of competition, rather than toward developing new capabilities relevant for the new basis of competition. For example marginal cost analysis makes RCS (e ,5) backboned on IMS is an attractive choice for new presence and messaging services designed according to traditional telco service models.

However, according to the new basis for competition, the scalability and interoperability offered by IMS are less important than flexibility. Telcos could be better off investing in new, more flexible infrastructure better suited for experimentation with new services, use cases and business models. As such the focus of Web RTC innovation should be on building developer ecosystems for voice services, discovery of new use cases and experimentation with new business models, and not on technology. If Telcos won't do this, competitors will.

Imagine for a moment that you run a mobile operator who derives 10% of its revenues from SMS (which commanded 70% margins in the heydays of mobiletelecom). In a matter of few short years, these pesky free messaging apps destroyed your most profitable business. At first you thought that these "free apps" would go away after they burned through the money provided by the venture capitalists. Then you tried to compete with homemade competitors like WAC and Joyn, only to discover that it's a sure recipe to lose money.Unfortunately, since the new generation of messaging apps have unique business models, they can sustain free services indefinitely and forever change the dynamics of the mobile messaging market. WeChat, Viber, Line and others monetise by using their platform for e-commerce, selling digital goods (stickers, games), physical goods and services (like taxi rides). They don't have to charge for the messages or even voice calls to be commercially successful.

Opportunities in digital content fall into two broad categories: production and distribution. In IBM's opinion, content production offers little potential for telecom providers; most operators will do better by partnering with content providers than by attempting to produce content themselves. However, they can also play a role in facilitating the trend toward user generated content by enabling consumers to enhance their own content with a range of telecom capabilities, including location, presence and interactive services. To defend and grow their share of the digital content market, Telcos will ultimately have to make a substantial organizational, cultural, technological, operational and business model transformation as they transition from providing network connectivity to enabling the consumer's digital experience. With the transition to Internet Protocol (IP) and the proliferation of content distribution platforms, consumers increasingly want choice, flexibility and control over the multi media multi screen experience. Telecom operators can draw on their unique skills and capabilities to capitalize on this trend and distinguish themselves from rival platform providers.

Vision Mobile states that Telcos must move their innovation focus from technologies (be it HTML5, NFC, IMS, VoLTE, M2M or RCS-e) to ecosystems. Ecosystems are much better at delivering choice and flexibility.Ecosystem economics are driven by network effects and lock-in. iPhone apps attract Apple users, who in turn attract more developers, who make more apps, which attract even more users, and so on. This network effect between developers and users drives the explosive growth of the iOS platform. Lock-in creates natural "walled gardens," as users develop habits around apps, while developers are locked-in by high switching costs created by their investments into the platform. A platform business model is about leveraging an operator's underutilised, walled network assets, taking a cut from the delivery of innovative services, in the same way that Apple takes a cut from the delivery of mobile apps or Facebook takes a cut out of ad delivery.

Over the longer term, Telcos can look for ways to build parallel ecosystems, using lessons from the ecosystem economics textbook. M2M holds the potential to create a vibrant ecosystem of users and solution providers, thereby establishing strong network effects and lock-in. Telcos can become the central force in this emerging ecosystem if they learn to engineer the ecosystems to their advantage. By looking at M2M through the lens of ecosystem economics, operators will see opportunities that are much bigger than just selling modems and data connections.But when it comes to making strategic decisions, the digital leaders have proventhat there is no reason to be bound by this artificial framing of markets.

The study of markets worked well for industrial age but in the Digital Era companies can get unfair but deserved advantage by smashing industry boundaries : That is by competing across multiple markets at the same time.Apple created an unfair advantage by competing in both consumer electronics and digital content markets. Google created an unfair advantage by competing in both online advertising and mobile markets. Amazon created an unfair advantage by competing in both e-commerce and tablet markets. Their direct competitors (Nokia, Yahoo, eBay) didn't stand a chance as the mobile revolution unfolded.

In the future, our communication will revolve around social media platforms, dominated by Internet firms such as Apple, Facebook and others that do not even exist yet. The Google+ video telephony project launched in 2011 shows the way forward for integrated communication environments that, building on a personal or organizational network, set up telephony, messaging, mail, chat or video links at the click of a button : Hey .....Customers don't buy access, they buy service ecosystems !!

--◆--

**Telco Digital # 6 :
The Agility Agenda**

Telco Global Connect

As the telecoms industry has come under increased pressure, through heightened competition from both OTT players and other telcos, many operators are attempting to transform in order to remain competitive. Typically these transformation efforts have straddled the line between cost cutting exercises (to streamline the business) and developing new techniques and tools to offer new services (to expand capabilities and offerings) to grow revenues. In an ideal world one would simultaneous transform senior management's mindset, strategy, structure, systems and culture. But in reality that's hard, if not impossible, to do. It is more realistic is to seek to change areas over time, building on successful initiatives. For example, an operator could make one business unit, process or system more agile, measure the results, and use the proof of success to lead senior management to embrace the concept.

Inefficient IT processes are an impediment for companies seeking to compete successfully against digital companies and improve business performance. As companies collect more customer information and need to transmit it in real time across applications, they require more storage and computational power. But rather than add more components to an already-complex system, companies would do well to pursue automation of what they already have. Indeed, the top-performing companies have pursued simplification and automation of various backbone IT processes and systems— for instance, automating server deployments, load balancing, and service-ticket management.

These changes not only have generated significant cost efficiencies for those companies but also have given the operators much greater flexibility in terms of service capacity and load volumes. In emerging markets where some fast-growing telecom operators are adding as many as one million to two million subscribers per month, the ability to automate capacity, server throughput, and storage has allowed senior managers to focus on business growth rather than scramble to augment their IT infrastructure.

Business agility is the "ability of a business system to rapidly respond to change by adapting its initial stable configuration" Business agility can be maintained by maintaining and adapting goods and services to meet customer demands, adjusting to the changes in a business environment and taking advantage of human resources.In a business context, agility is the ability of an organization to rapidly adapt to market and environmental changes in productive and cost-effective ways. The agile enterprise is an extension of this concept, referring to an organization that utilizes key principles of complex adaptive systems and complexity science to achieve success. One can say that business agility is the outcome of Organizational intelligence.

Similarly, agile enterprises do not adhere to the concept of sustained competitive advantage that typifies the bureaucratic organization. Operating in hypercompetitive, continuously changing markets, agile enterprises pursue a series of temporary competitive advantages—capitalizing for a time on the strength of an idea, product, or service then readily discarding it when no longer tenable. Enterprise architecture as a discipline supports business agility through a wealth of techniques, including layering, separation of concerns, architecture frameworks, and the separation of dynamic and stable components.

BSS systems (typically including billing and CRM), have always been separate from OSS systems (such as resource management, service activation, provisioning, fault management, etc.), which included having separate business processes and people. For example, revenue focused BSS was always run by the IT department, and cost-focused OSS was run by network operations. This traditional binary approach would have likely continued to be sufficient if not for the major transformation the telecommunications industry is undergoing, where service providers are becoming retailers of multimedia and entertainment services.

Over the years Telco IT infrastructures have evolved into expensive, complex collections of monolithic applications interconnected with specially built point-to-point interfaces. Transforming OSS/BSS platforms and partially rebuilding with more modern and harmonized platform components leads to considerable savings in the long run: reductions in time, effort and costs spent on system integration, administration, maintenance and training. An enduring vision of harmonized OSS architecture is inspired by the TMF Lean Operator Initiative and based on four key areas:

• CSP's process architecture: The CSPs' business processes can be supported by introducing modifiable operator process templates out-of-the-box and enabling a higher level of automation in their daily routines.

• Common information architecture: stepping away from "stove piped" data and supporting shared information and data models. This enables OSS/BSS level application interoperability through Common Information Models.

• Modular application architecture: will bridge the gap between service and resource management applications. A high level of modularity allows flexible solution building: it enables easier maintenance, allows changes on one component without affecting others and allows new components to be added as required.

• Application integration architecture: Interoperability and time to market is improved through compatible interfaces, common information models and through leveraging partner ecosystems and productized adaptation libraries.

Bear in mind that the transformation to the next generation OSS is a revolution, nor is it fixed to a particular date or year. As current OSS systems are crucial for the operation of a network they cannot be replaced overnight. The transformation and migration will need to happen gradually, making the challenge even greater – old systems cannot be turned off before new systems are in place. To mitigate these risks, future needs must be anticipated in advance and OSS architecture must be designed to fit with future requirements from the start. OPEX for the legacy OSS needs to be reduced to make room for new investments and replacement of the old functionality. OPEX reduction takes many forms, including:

• Removal of old, redundant OSS applications and systems
• Streamlining of functionality in legacy OSS
• Replacement of bespoke/ customized systems integration work with standards-based software and off-the shelf mediations
• Selective freezing of legacy OSS applications and systems
• Encapsulation of functionality and making it "service aware" with SOA
• Effective use of key OSS systems, moving functionality to these and taking other systems off-line

The Enterprise Architecture Model describes the elements of business – strategy, business cases, business models, processes, supporting technologies, policies, and infrastructures that make up an enterprise. It also provides means for governing the enterprise and its information systems, and planning changes to improve the integrity and flexibility. In other words, Enterprise Architecture crystallizes the organization – what it has to do and how – to be as efficient and productive as possible.

In the Enterprise Architecture, the business quadrant handles the value chain aspects relevant to the business as a whole: where to improve the business efficiency and develop new value propositions and how to increase efficiency and competitiveness of the business in the context of its environment: markets, competitors, legislative and environmental aspects, influences and impacts.With Enterprise Architecture (EA), new opportunities and capabilities will raise some real competitive advantages for Telco operators. Architecting the future state of EA is the heart of the entire process. The goal is to translate business strategy into a set of prescriptive guidance to be used by the organization (business and IT) in projects that implement change. As such EA is a process discipline. Done well, it becomes an institutionalized part of how an organization makes decisions to direct its investments, such that the chosen business strategy will be realized. Usually a system is seen as a necessary cost to make the business – not anymore and certainly not with EA !!

Managing IT complexity to support business strategy is a big challenge for enterprise architects at large companies when a company has global operations, as is the case for Telstra, an Asia-based telecommunications firm. However Telstra's enterprise architecture (EA) team addressed its challenges by focusing on customer engagement, improved agility, and global business strategy enablement.The EA process bridges the gap that otherwise exists between business strategy and technology implementation. High-performing organizations are process-disciplined which is lacking in many of the Tier 2 Telco operators.

In turn, every high-performing process must be defined/documented, have process owners and be closed-loop with governance in place. Activities in this phase of the architecture process include but not limited to :

• Scoping the EA program and the next iteration thereof in terms of breadth and depth, which is known as defining what is meant by "enterprise"
• Gaining executive sponsorship and support
• Conducting stakeholder analysis
• Identifying the EA leader or chief architect
• Building and chartering the "EA team," which will own and facilitate the EA process and establishing clear roles and responsibilities
• Assessing organizational readiness and EA maturity
• Developing an initial communications plan, communicating the role of EA and setting expectations of individuals participating in the process
• Establishing a plan for setting up a governance mechanism
• Defining measures of success to articulate value delivered

Currently many Telcos are burdened by a wide range of systems 'isolated' for the operation of its business. This reality does not allow effective sharing of information between systems and / or applications. In recent years they have acquired several technologies were acquired from different manufacturers and suppliers, most could be considered islands of information and technologies. Today's service providers must close the gap between their Customer-facing BSS and network-facing OSS. With Enterprise Architecture (EA), new opportunities and capabilities will raise some real competitive advantages. As an example, let's consider a set of typical (separated) systems:

• Automation system: The principle behind this is to improve efficiency, automating several steps (or all steps) of certain tasks. Since the tasks are automatic, the delay is caused by latency of the system itself, leading to execution of thousands of tasks per second instead of seconds/minutes spent in each task.
• Customer segmentation: Groups people according to attributes that store information relevant for understanding customer behavior, and can be used to predict the probability of acceptance or refusal of a certain product or probable churn.
• CRM tool: Contains all customer information and supports the call center team in customer interactions.

The new reality in the Telco industry is that the basic currency of the smart network is DATA. The move to all-IP networks and the technology that has become available means that operators can collect more data than ever before from all points between their core networks and their end users and exploit it in ways not previously imaginable. Excellence in IT architecture is fundamental to efficiency and effectiveness, touching every aspect of a telco's business performance. Fortunately most senior Telco execs have already realised that the integration of information systems, collecting, consolidating and making available all data efficiently is an essential requirement to ensure the viability and competitiveness, avoid errors and waste, improve efficiency and increase the success factors internal. As such any strategic plan to transform the BSS/OSS using EA must :

- Align the needs of information systems with business strategy,
- Monitor the rapid evolution of Information Systems,
- Rationalize and monetize investments in Information Systems
- Prioritize solutions to develop in the future according to the business strategy defined by the company
- Controlling the proliferation of systems / applications isolated and walk to the integration and overall management of Information Systems

The IT industry has embraced the concept of a Service-Oriented Architecture (SOA) as a standardized, more efficient way to build enterprise IT infrastructures. I believe that SOA, together with a revised enterprise business process, is the right way to build BSS and OSS applications, because it supports more agile internal operations, enables interoperability among new applications, and can be used to leverage existing BSS and OSS assets by adapting them to the SOA model. However to yield genuine value, an architecture transformation also requires a substantial shift in mindset. It is crucial to nurture the partnership between business and IT, rather than allowing IT departments to function in their default mode as delivery organizations and service providers.

Transforming a large telco's enterprise architecture management function to deliver maximum value is a Herculean task and multi-year effort requiring full buy-in from the business side. There is no uniform panacea for success. But the impact on costs and business performance can be huge once the enterprise architecture moves toward a uniform blueprint with consistent management across domains. Tariff changes take days rather than months. Customers can be tracked across their lifecycle and targeted with optimally customized offers, while network utilization soars.

Globe Telecom Inc. is a leading telecoms company in the Philippines, a country with a population of more than 98 million people. The company's mission is to enrich and simplify everyday communications to bring customers closer to what matters to them most; its vision is to have the "happiest customers".Globe's journey toward achieving these goals started four years ago and included formation of the company's Enterprise Architecture Division, which is tasked with establishing the right structure, framework and governance in its architectural planning. The company chose TOGAF (see below) as the basis for its enterprise architecture and leveraged TM forum's Frameworx suite of models, tools, best practices and standards to implement it.

Using enterprise architecture blueprints, Globe was able to define strategies and develop target architectures to transform its business and operational support systems (BSS/OSS). The blueprints have also helped Globe identify gaps in the architecture and set up initiatives to address them in the application architecture, for purposes beyond the BSS/OSS transformation, to gain new customer-supporting functionality.
Some of the new functionality is already in place, while some is still in the planning stages:

- New digital services – the digital media and commerce ecosystem is part of Globe's drive to create new revenue streams as core revenue erodes. This is through partnerships with over-the-top companies providing apps and digital content, including established favorites such as ringtones. Although still in the early stages, the ecosystem handles payment settlements and revenue-sharing between partners. The service delivery platform enables the digital media and commerce ecosystem; it handles which application program interfaces to expose and how, to whom, to provide and deliver digital services.

- Understanding customers – Globe is hoping unified user profiles will help the company build a 360-degree view of its customers, adding new sources of data to the usual data operators hold about their customers, such as where they are and what they are doing on their devices. The relevant data would be processed in real time or near real time to constantly update the profiles and make them as accurate and complete as possible. Over time, this would give Globe an ever greater understanding of its customers and help the company to target promotions and services.

In addition, contact policy management would allow customers to manage their preferences regarding how (which channels) and how often Globe contacts them. Customers would be asked to opt in, rather than the onus being on them to opt out.

- Measuring and managing customers' experience – the company is exploring different ways of measuring customer experience and the key performance indicators (KPIs), including in the core network. Uses for this information include being able to identify clusters of high-value customers and address any issues they have regarding quality of service, as well as identify problems and the best solutions to them.

- Pioneering uses of analytics – Globe's use of analytics goes way beyond these more traditional uses, for example, by incorporating unstructured data from social media – and responding to customers using social media as additional channels. The company also incorporates data from partner organizations to identify customers and round out their profiles to gain a better understanding of their preferences, usage, habits and concerns, and it is exploring development of another revenue stream by aggregating and anonymizing this consolidated data to sell to other organizations.

Another strand of Globe's analytics activities is master data management, which includes centralizing access to non-transactional data entities like customers, products and partners, resource data, and reference data such as postal addresses and street names, among other things, to ensure that data is high quality and consistent in definition and usage across the entire company.

- Better processes, greater efficiency – Globe has deployed an enterprise-wide business process management engine to automate and streamline processes across all its domains. This allows disparate applications to participate in an end-to-end process via the business process management architecture, regardless of where they sit in the organization.

This has dramatically reduced the levels of fallout from processes and provides a holistic view of them. Work is underway right across the company to fully map all processes to the Business Process Framework, so that everyone can operate via the business process management engine. The enterprise service bus gives Globe the capacity to add internal applications and communications using the service-oriented architecture approach. This limits the impact of additions on other parts of the organization.

The standardized approach enabled Globe to use commercial-off-the-shelf products to keep costs down and ease ongoing integration. It also means the applications are portable, giving the company many choices about whether to outsource them or run them in the cloud to gain greater cost and operational efficiencies.

- Controlling internal content and additions – the enterprise content management facility is another facet of the business process management engine. It handles and stores digitized documents such as contracts with customers, partners and suppliers, and enables version control to reduce instances of mismatched data and order fallouts, for example.
- Globe's new architecture has reduced the level of complexity in IT and network infrastructure, through greater independence of applications and services, rather than them being tied to discrete systems. Hence Globe has unprecedented flexibility to build, buy or outsource IT and networking solutions.
- This has removed the mass of point-to-point linkages and given the company greater ability to address critical, enterprise-wide issues like security and privacy though common repeatable processes and applications being able to communicate with others as necessary.
- Simplified architecture and processes clarifies who owns and is responsible for what, including who makes final decisions. Previously the more siloed approach meant no one had ownership of a process from end to end.
- The cross-enterprise architecture and model enables better interoperability, simpler system and network management, and easier upgrades and changes to existing components.

In short, the company is benefiting from much better enterprise-wide alignment in all aspects between the business, IT & networking organizations, including in strategy, processes, architectures and governance.Most operators have a long way to go to become more agile as an organisation and to embed news ways of working. In order to achieve this vision STL recommends that operators:

- Secure Senior Management Support – Senior Management must fully embrace the need to change entrenched cultures and processes and adopt a more agile mindset and approach.
- Focus on the Customer – An operator's focus should shift from their products towards the needs and behaviour of the customer. This is achieved through both

a mindset switch and organisational structure change as well as embracing relevant metrics, tools and technologies.

- Simplify Processes – In order to compete against OTT players (& other telcos) operators need to move at much faster speeds. Processes therefore need to be (relatively) simple, enabling decision-makers to act quickly.
- Focus on Innovation – Senior Management should prioritise and encourage innovation across all teams. Operators should regularly undertake new initiatives and aim to quickly identify initiatives that will not succeed and learn from them, embracing the 'fail fast, fail well' mantra.
- Embrace Agility-Enabling Technologies – Technology can further increase an operator's ability to move at speed and to innovate.

Analytics technology can help operators better understand service performance/usage and customer needs, helping operators to refine and improve their offering .Service creation tools enable operators to create new products and services more rapidly Similarly, flexible OSS/BSS systems allow operators to manage new (& existing) services more effectively and to track associated billing requirements. Enterprise Mobility tools allow employees to work more flexibly, reinforcing an agile culture. XaaS models can potentially free up capital and staff costs (allowing operators to focus resources on other areas), and it can reduce the risk when trying something new

Operators need to regularly take stock of changes in the market and adapt accordingly to ensure continued relevance in the future. Operators should develop new metrics to both track market developments and to assess their ability to react and evolve to meet the changing market environment. Senior Management should leverage this information to review the organisation's strategic direction.

Today's mobile networks are limited and built upon a best-effort design, but that means they have latency issues and cannot dedicate high bandwidth to a particular user on the fly. Network virtualisation highlights the transformational path that operators are willing to take to counter the stress that financial pressures are putting on profitability while effectively and efficiently monetising data growth and reducing vendor lock-in. This trend clearly shows that, in order to be sustainable in the near-future, operators networks will require the right amount of mobility, ultra high-speed networks, cloud computing, big data.

Network virtualisation allows operators to simulate network resources through SDN and NFV technologies that decouple, run and optimise different functions of the network.The industry is evolving from proprietary equipment networks to IT-based data centre networks that employ technologies such as software-defined networking (SDN), network function virtualisation (NFV), cloud-computing and big data analytics to provide a variety of converged services to consumers. NFV is highly complementary to SDN. Network functions can be virtualised and deployed without an SDN being required and vice-versa. According to ETSI, early NFV deployments are already getting underway and are expected to accelerate during 2014-15.

Software-defined networking (SDN) and Network Functions Visualization (NFV) will drive changes in data security investment, according to a new report from Infonetics Research. Their Data Center Security Products report noted a shift in how organizations protect digital properties, including a 44 percent rise in the sale of purpose-built virtual security appliances. They anticipate a fairly significant revenue transition from hardware appliances to virtual appliances and purpose-built security solutions that interface directly with hypervisors, with SDN controllers via APIs, or orchestration platforms.

Rather surprisingly, communications service providers (CSPs) themselves, not vendors, are driving the development of network virtualization technologies. The potential to dramatically accelerate new service delivery, lower operating costs, and eliminate vendor lock-in has CSPs salivating and network equipment vendors scrambling. Vendors who sell proprietary network gear don't exactly welcome the thought of their intellectual property being replaced by standardized software running on commodity hardware.

This has pushed the timeline for SDN and NFV further out, and prompted more than a few analysts to pull the hype card.The virtualization of service and control functions in the core network has been a first step in using cloud computing technology in the telco domain. However, for a full telco cloud implementation, virtualization needs to be complemented with a complete cloud platform and management system. This must include classical network management for legacy systems, plus virtualized network function, cloud orchestration and application management to achieve the full benefits of automated provisioning and elastic scaling of the network.

Driven by the promise of total cost of ownership reduction, wireless carriers are aggressively jumping on the NFV and SDN bandwagon, targeting integration across a multitude of areas including radio access network, mobile core, OSS/BSS, backhaul, and CPE/home environment.Telecom Italia has been among the tier 1 telcos driving the move to NFV. Along with ATT, BT Group, Deutsche Telekom, Orange, Telefonica and Verizon, the company a couple years ago pushed network functions virtualization into the spotlight by creating an ETSI group to explore the technology. The key goals of the NFV Working Group are to reduce equipment costs and power consumption, improve time to market, enable the availability of multiple applications on a single network appliance with the multi-version and multi-tenancy capabilities, and encourage a more dynamic ecosystem through the development and use of software-only solutions.

Telefonica's UNICA platform is initially focused on virtualising signaling-related functions, including IMS (IP multimedia sub-system, DNS (domain name system), SMSC (short message service centre) and OCS (online charging system). The second phase will look at virtualising functions that carry traffic such as the core packet network. Telefonica's NFV programme is notably designed to "source different functions to different suppliers" and avoid vendor lock-ins. The company wants to design a virtualised network architecture that allows vendor interoperability.Among the many capabilities offered by UNICA is the idea of multi-tenancy (where the same basic solution works for multiple organisations) or NaaS (Network as a Service), using pre-installed templates to deploy virtualised equipment in real time and with integrated resource management.UNICA promises to offer real and permanent change for Telefónica's network transforming the company into a true Digital Telco.

Meanwhile ATT, has introduced its vision for the company's network of the future: the 'User- Defined Network Cloud.' AT&T claims their the cloud-based architecture is "a global first at this scale." The operator also announced the group of vendors that will work on implementing this strategy. The carrier expects its revamped architecture will accelerate time-to-market for technologically advanced products and services. Integrated through ATT's wide-area network (WAN) and using NFV and SDN, the architecture is expected to simplify and scale AT&T's network by separating hardware and software functionality, separating network control plane and forwarding planes, and improving functionality management in the software layer.This move to software-based telco environments will not only help incumbent providers become more agile and adapt to market trends and subscriber demands more effectively, but will open up the market to new players who may not have had such deep pockets needed to develop proprietary hardware. It will allow new carriers to quickly scale and compete, as they won't have to load up on costly central office equipment to get started.

Traditional telco business served a well-defined need that consisted in person to person remote communications which later extended to machine communication. Today we are in a completely different universe. Digital ecosystems have radically shifted the well-established paradigms enabling endless possibilities, adding extra layers of richness and complexity and dramatically accelerating products timescale. To create new digital telco businesses in a turbulent context we have to lead and build a sophisticated capacity for continuous learning. It is people who learn and lead no one else, so the only way is to unleash people's talent. The organization has to evolve from hierarchical, verticalised and process centred to talent-driven, where people take the responsibility and are empowered to propose, defend and execute innovation projects with autonomy.

Lean startup is a method for developing businesses and products first proposed in 2008 by Eric Ries. Based on his previous experience working in several U.S. startups, Ries claims that startups can shorten their product development cycles by adopting a combination of business-hypothesis-driven experimentation, iterative product releases, and what he calls validated learning. Ries' overall claim is that if startups invest their time into iteratively building products or services to meet the needs of early customers, they can reduce the market risks and sidestep the need for large amounts of initial project funding and expensive product launches and failures.

The lean startup philosophy is based on lean manufacturing, the streamlined production philosophy pioneered by Taichi Ohno by combining flow principles used by Henry Ford starting in 1906 and the TWI programs introduced to Japan in 1951. After more than 15 years of experiments, he had a stable and reproducible system. Use of the term *lean* to describe Ohno's system was first formalized in the book *The Machine That Changed the World*. The lean manufacturing system considers as waste the expenditure of resources for any goal other than the creation of value for the end customer, and continually seeks ways to eliminate such waste.

Similar to the precepts of lean management, Ries' lean startup philosophy seeks to eliminate wasteful practices and increase value-producing practices during the product development phase so that startups can have a better chance of success without requiring large amounts of outside funding, elaborate business plans, or the perfect product.Ries believes that customer feedback during product development is integral to the lean startup process, and ensures that the producer does not invest time designing features or services that consumers do not want. This is done primarily through two processes, using key performance indicators and a continuous deployment process.

Because startups typically cannot afford to have their entire investment depend upon the success of one single product launch, Ries maintains that by releasing a minimum viable product that is not yet finalized, the company can then make use of customer feedback to help further tailor their product to the specific needs of its customers.Based on Lean Startup the three reference pillars of the framework Telefonica applied to its innovation are:

1. Start small and aim high: the level of ambition in our innovation projects must be high. They need to bring the possibility of a global reach and the potential to make an impact in everyday life and business. This does not mean that they will burn lots of resources to start with or that they need to show full potential from day one, quite the contrary. Projects, specially at the beginning, work just with bare minimum resources, and then investment increases as the project progresses with validated learnings. The less uncertainty the more budget.
2. Iterate fast to achieve efficiency in each of the maturation stages. This means scaling down the initiatives that are too early in time, inmature or unfocused while fueling up the ones that show traction. Therefore, product investment decisions conducted along the process rely not only on technological trends, but also on a profound understanding on which markets the digital customers will form in the upcoming years.
3. Fail fast, fail cheap and make sure you learn along the way. Instead of devoting large quantities of energy and resources to increase the individual success chance of a few projects we've thought it wiser to lower the overall risk by minimizing the failure cost for each project.

Telefonica has realised that Lean Startup has allowed us to accelerate the innovation cycles, through many short iterations within the projects. With a more traditional methodology innovation projects would mainly focus on getting working prototypes which aligned just with the initial specifications in a timeframe of around 4 months. Only after the initial prototype was available the team would start thinking about how to include the development in the roadmap or how to add it to an existing product. Reality checks with the market in general were not part of the innovation process itself that would usually only happen at the end of the project. The innovation projects which apply Lean Startup start generating meaningful learnings in 1.5 months or less. They have estimated this is an acceleration factor of 2.6. Which means they can figure out if a project/products makes sense at least 2.6 times sooner than before.

Telcos have gone through several waves of transformations in the past two decades, in a way which is probably unequaled in any other industry, and this trend will continue for the foreseeable future. Beyond what the industry has already done, it now has to – like several other sectors – undergo radical digital transformation, which is basically three-fold: To deliver an outstanding and differentiating digital customer experience: To acquire a strong and profitable position in the digital services value chain Operations and To develop efficiency and agility.

In addition, as innovations in the customer experience and business models areas can be quickly copied and pasted by others, digital transformation of operations remains more confidential and can therefore be an even stronger and longer-lasting differentiator. As for all industries and alike all radical transformation initiatives, moving to digital operations is a tough and risky journey, with some specific challenges for telcos, such as:

- The high complexity of their infrastructures, both IT and network. Getting agile in this context is like having to maneuver a big ship with agility
- The need for a highly safe agility, because of their critical role in the cybersecurity chain
- The conjunction with other structural changes in their environment, including new networks technologies (SDN, NFV, 5G,...) and global consolidation of the industry.

As convergence unfolds, a performance culture could transform the earnings paradigm. Certain telecommunications companies have applied vision, daring, and discipline that have led to adding a percentage point per annum to their operating profit during the past decade, despite some historically adverse economic cycles, while logging comparably solid sales growth in relatively mature, slow-growth market sectors. Convergence could compound growth differentials in both sales and earnings dramatically. Telecom companies with consumer value cultures – that is, with strategic agility and execution discipline coupled with strong innovation leadership – will then be positioned to grow their market capitalization outpacing their industry peers

For many operators in MEA Region, the cost of delivering services will increase faster than revenues are keeping pace. In fact, this so-called revenue-cost gap is generally a global phenomenon, as operators find that single digit percentage reductions in expenses are simply not enough when it comes to meeting customer expectations for personalized multimedia services. As such the entire concept of "managed services" has changed.The complexities of running both fixed and mobile networks and value added services continues to increase, while operators are increasingly required to focus their scarce resources given a highly competitive market. To maintain margin growth operations have focused on cost-cutting exercises which has accelerated outsourcing.

Operator's acceptance of partnering and acquisition of external expertise has grown in recognition of the opportunities and efficiencies such activities provide. Managed services has emerged as an important delivery model for network equipment vendors, systems integrators, BOSS specialists, applications and service specialist as well as network-owning operators themselves.The main components of the old approach — NOC services, multivendor maintenance and network outsourcing — were sufficient in an environment where point solutions, cost savings and the bottom line were primary considerations. According to Analysts many companies are undercutting their own outsourcing initiatives by not managing them strategically. Most outsourcing initiatives do not attain the full desired benefits, especially cost reductions simply because Telcos fail to :

• Define a clear strategy of what activities should be outsourced, what objectives should be accomplished via outsourcing, and what boundaries should exist between any internal and external work. Aligning this strategy with the business units to maximize the value of any sourcing decisions is critical
• Establish a clear baseline of current costs (both the total and its components) and compare that to external benchmarks. Deciding what work can be optimized internally prior to outsourcing and use that as the basis to define cost targets with vendors is paramount
• Develop to establish a systematic demand management process that prioritizes project requests based on the proposed value creation for a business. And introduce consumption-based chargeback mechanisms to the business units to create an awareness of costs and to foster behaviors that are more conscientious toward usage.

Buyer organizations should use a structured, strategic approach to calculate the true costs of outsourcing, taking into account the risks and opportunities. If cost is a concern, then a clear knowledge of performance baselines before entering an outsourcing engagement is critical. To contain costs, companies should rely on benchmarks and adopt a consumption-based system that charges expenses back to the business units that incurred them.Providers and buyers must work together to apply sourcing discipline and craft outsourcing relationships that address near-term cost objectives and longer-term scalability and enhancement. Clearly companies need to turn transactional vendor relationships into strategic partnerships by doing the following:

• Clearly define the objectives and expected results (KPIs) from any outsourcing partnership.
• Assign clear roles and responsibilities to in-house personnel for handling various aspects of vendor partnerships, from overall relationship management to individual projects and their specific deliverables.
• Agree on an interaction model with the vendor to ensure day-to-day communications occur, expectations are regularly aligned, potential problems are identified and addressed early on, and future opportunities, especially those involving new technologies, are identified.
• Flexibility is the operative word for outsourcing strategies and contracts because outsourcing can never be separated from business goals. To drive the desired results, outsourcing relationships must constantly evolve.

In the traditional BOM approach, a qualified and trusted partner can share a portion of the risk in the early years of deployment with the network operator. Instead of paying all CAPEX up front before realizing any new revenues, the network operator can share the risk of build-out with the partner by deferring a share of the cost until it realizes subscriber growth.Since next-generation managed services are geared toward achieving business transformation via services-centric approach risk mitigation is evolving into a risk sharing approach. Risk avoidance will be an inhibitor for some companies to outsource, even those that could gain efficiencies from managed services and performance-based contracts with an external provider. Risk management best practice include a PLAN :

• To identify and evaluate technical and commercial risks due to outsourcing specific services.
• To take mitigating precautions wherever warranted, such as identifying alternate vendors for crucial pieces of work.
• To define and continuously monitor risk-related KPIs that trigger early action whenever a particular risk materializes.

Customers should understand that asking suppliers to accept a greater proportion of risk will have a direct impact on the level of charges payable under the contract. For example, the transfer to the supplier of risk which would be best managed by the customer will lead to an increase in charges.Customers need to take a realistic approach to risk allocation; there are some risks which cannot simply be "outsourced" to suppliers. Indeed, attempting to do so could damage their core business and potentially put them in breach of their regulatory obligations. In negotiating contracts, customers should focus on those risks which are real and important to their business. Where the risk is not important, they should take a pragmatic and commercial view in negotiations.

Although buyers expect service delivery performance to match what has been contractually agreed on, as the focus shifts increasingly to cost reduction, providers and clients must be able to manage the relationship and the expectations, not just the contract. Over the lifetime of the deal the focus is likely to shift rapidly toward improvement and innovation. The challenge will be in managing change while maintaining realistic expectations among the key deal stakeholders within the context of the deal.

To gain advantage from the major trends driving the rapid evolution of infrastructure outsourcing, buyers must accept — and when possible seek — opportunities for consolidation, industrialization and global delivery. The best outsourcing contracts ensure the proverbial win-win relationship, where both parties arrive at terms and pricing that is fair and promises long-term sustainable value. Contracts that leave the provider with only marginal profit or limited revenue growth will ultimately result in service quality issues.

In setting up these managed services contracts, operators' main goal has been improved cost efficiency — a result of the outsourcing vendor's ability to realize economies of scale, and the global knowledge and experience of the delivery partner. In general, the decision to contract for managed services is made at operating-company level based on local circumstances. While managed services contracts have cost efficiency built in, operators still need to see hard evidence of major cost advantages before they will be convinced to outsource more activities according to Ey.

There is fast growing trend toward sale and leaseback of tower infrastructure certainly in Africa. Telcos use of this technique varies on a regional basis, reflecting a lack of infrastructure players in some markets, and the fact that some regulatory authorities do not allow it. There is also a degree of uncertainty over the extent of the benefits to operators — and over whether the tower companies will share the same objectives as the operators in the future. In tower transactions, the contract and counterparties involved vary from an outright sale to a third-party infrastructure provider, to a transfer of assets to a tower company created and wholly owned by the operator itself, either on a stand-alone basis or in a joint venture.

Selecting the most appropriate supplier for a project can reduce risk substantially. It is incumbent on a customer to devise a due diligence process that will properly test and evaluate potential suppliers. A successful due diligence exercise should not just be a paper exercise; it should involve visiting potential suppliers, testing technology and speaking to other customers of the supplier. It is also important for customers to consider soft issues such as cultural fit. All too often outsourcing goes wrong because it was not possible to create an effective working partnership between customer and supplier.

It is clear that whilst the general economic situation remains difficult, customers will continue to be under pressure to reduce their costs and will look to their suppliers to help them do so. When it comes to entering into new arrangements customers should draft their contracts to allow for the maximum possible flexibility. No business or business environment remains static – change is inevitable. The contract should therefore contain a mechanism for managing contractual and operational change. Once again, good governance is key.

Through creative partnering and innovative risk sharing options, new managed services and outsourcing business model options provide the framework for creating a next generation enabled portfolio of services for consumers and enterprises ready for Telco 2 sided business model. Instead of short-term tactical advantages, the focus is firmly on long-term strategic gains . And only then will Telcos be able to reap the full benefits of managed services via trusted partnerships.

---♠---

Telco Digital # 7 : The Leadership Agenda

**Telco Digital # 7 :
The Leadeship Agenda**

Agenda!

Telco Global Connect

Telcos have a number of attributes that give them a potential marketplace advantage in the Digital world of commerce and communications : an extensive customer base, distribution muscle , knowledge of customer preferences through CRM and billing systems and hopefullt an elightened leadership team. Perhaps the best way to start is understanding this digital animal is by highlighting what a digital telco's objectives and then work backwards to know what needs to be done.

- Digital Telcos aim to provide the digital products and services which will help to improve the lives of our customers by leveraging the power of technology. This ranges from developing new technologies for consumers to communicate with friends and family through to helping businesses and governments address new opportunities, improve operations and increase efficiencies.
- Digital Telcos implement a new organizational structures that is completely focused on clients and incorporates this digital offering as the main focus for commercial policies. The structure gives greater visibility to local operations, bringing them closer to the corporate decision-making centre, simplifying the global structure and strengthening the transverse areas to improve flexibility and agility in decision makings.
- Digital Telcos deploy a new breed of IT-centric services that involve the use of data, analytics and digital content. CIOs transform business-only, support IT to effectively deliver customer-facing digital services.To transform CSPs into diversified service providers CIOs harvest business value from existing networks, IT and information assets, and hunt for new opportunities using digital services.
- Digital Telcos are geared to deliver high quality, integrated product offerings across a number of more established digital service areas, including Machine-to-Machine, Cloud Computing, eHealth, Financial Services, Advertising and Information Security. It also brings together all capabilities in media services ranging from IPTV and satellite TV to over-the-top video, ensuring that the Digital Telco is a leading video broadcasting company as well
- The transformational journey to Digital Telco allows operators to gain an intimate understanding of their customers, become more responsive, improve satisfaction and address diverse needs on a personalized basis. The digital transformation needs to be a strategic goal and come as a result of a real passion to change the operator's role in the digital lifestyle.

Big Data Analytics related services are also a natural fit in the Telco digital services offerings landscape, considering their technical prowess and historical experience in analysing customer data. Telcos have traditionally exploited customer data to focus on churn reduction, design tariff plans and manage campaigns, and are well placed to leverage Big Data as a new source of revenue.

Now lets get to the real challenges of going Digital... the part that overwhelms well intentioned Telco execs. Legacy systems present technical challenges when attempting to change, modify or expand offerings. This constraint both dramatically increases the time it takes to launch or modify services and also limits what operators can create and deliver. Cultural barriers prevent operators from moving fast enough. With a rigid, functional organisational structure, no-one in particular has a full overview of the whole (customer) process and they are therefore not able to understand the different dimensions, constraints, usage and experience of the product/service.

As such your digital ambition must be scrutinised by asking some difficult questions. How ambitious are your goals and what is the strength of your budget ? What do you plan to focus on when rethinking the customer experience? Are they ready to disrupt their entire business model, perhaps expanding into unknown territory? What is the state of your I.T Core ? How will you dovetail existing transformation and rengineering processes into a Digitised Telco ? Whatever the case, the key is determining how they can differentiate the company through digitization, whether in customer experience, with digital products and services, by becoming a trusted digital partner, or something entirely different.

Going digital will oblige you to appoint a champion reporting direct to the CEO...yes without CEO backing you might as well continue impressing your industry peers at the coffee table about how clever you are. Going Digital is a result of changing the mindset, operational structure and processes as well as transforming the legacy IT stack. The Digital Champion must practice Tipping Point leadership style so that newly Digitised Operator understands the potential dynamic shifts in customer usage behaviour and calibrates his Organisation to take both tactical and strategic actions to combat or align with the new digital world.

The conventional theory of organizational change rests on transforming the mass. So change efforts are focused on moving the mass, requiring steep resources and long time frames — luxuries few executives can afford. Tipping point leadership developed by W. Chan Kim and Renée Mauborgne, by contrast, takes a reverse course. To change the mass it focuses on transforming the extremes: the people, acts, and activities that exercise a disproportionate influence on performance. By transforming the extremes, tipping point leaders are able to change the core fast and at low cost to execute their new strategy.

Hence, contrary to conventional wisdom, mounting a massive challenge is not about putting forth an equally massive response where performance gains are achieved by proportional investments in time and resources. Rather, it is about conserving resources and cutting time by focusing on identifying and then leveraging the factors of disproportionate influence in an organization.

By single-mindedly focusing on points of disproportionate influence, tipping point leadership helps managers topple the four hurdles to strategy execution quickly and at a low cost by answering the following questions:

- What factors or acts exercise a disproportionately positive influence on breaking the status quo?
- On getting the maximum bang out of each buck of resources?
- On motivating key players to aggressively move forward with change?
- And on knocking down political roadblocks that often trip up even the best strategies?

To varying degrees, companies may face four types of hurdles to strategy execution. Knowing how to triumph over these organizational hurdles is key to successful strategy execution. Four hurdles to strategy execution were identified in the Blue Ocean Leadership paradigm :

- The Cognitive Hurdle: Waking employees up to the need for a strategic shift. Red oceans may not be the paths to future profitable growth, but they may have served the organization well historically, so why rock the boat?
- The Resource Hurdle: It is assumed that the greater the shift in strategy, the greater the resources it requires for execution.
- The Motivational Hurdle: How do you motivate key players to move fast and tenaciously to carry out a break from the status quo?
- The Political Hurdle: As one manager put it, "In our organization you get shot down before you stand up."

The theory of tipping points, which has its roots in epidemiology, is well known; it hinges on the insight that in any organization, once the beliefs and energies of a critical mass of people are engaged, conversion to a new idea will spread like an epidemic, bringing about fundamental change very quickly. The theory suggests that such a movement can be unleashed only by agents who make unforgettable and unarguable calls for change, who concentrate their resources on what really matters, who mobilize the commitment of the organization's key players, and who succeed in silencing the most vocal naysayers.

According to Egon Zehnder , to effectively target end users and create new digital value added services, Telecom executives must optimize 2 fundamental, highly interdependent skill sets :

Diversity of expertise : Telecom professionals tend to specialize very early on, choosing, for example, connectivity (wireless or wire-line), core transport technology, or, more recently, software development. As a result, individuals who possess a deep understanding of multiple engineering disciplines – as well as cloud based services, and the like – are extremely rare. And with every part of the technology landscape in flux, no individual can keep pace with all that is applicable, valuable, and relevant. That makes diversity of expertise, experience and perspective imperative. To fully leverage cutting-edge technology, telecom executives must forge cohesive, high-performing organizations that effectively blend diverse talent from multiple disciplines and backgrounds.

Companies are pursuing the creation of high-per forming organizations through a number of avenues. For example, some telecom companies are now realizing the power of the brand as the way to access the consumer, grow business, and ultimately add value in the market. They have recognized branding, marketing, and the ability to change customer perceptions and win mind-share must become core competencies of the organization. More and more companies are focusing on researching trends, segmentation, and consumer insights as a way to grow the business – redefining the role of the Chief Marketing Officer (CMO).

Today's CMOs must thoroughly understand analytics, data mining, and cloud computing. Through a strong liaison with the Chief Information Officer, they must tap into customer trends and be able to follow them in real time, as they develop. And they must use all traditional and non-traditional promotion and campaign channels effectively and simultaneously, from public relations to event marketing, online, and mobile. In this vein, several telecom companies are achieving even greater diversity of experience by entrusting their futures to executives who – though lacking specific experience in telecom or sometimes even in the technology sector – are proven visionaries, magnets for great technology talent, and inspirational leaders. Clearly, some are already moving to compete in a fast-changing world.

Cultural sensitivity : To integrate diverse talent, telecom companies must evolve toward more open, collaborative cultures that intrinsically value creative freedom, dissimilar perspectives, and far-reaching innovation as much as they have traditionally valued predictability and conformance to specifications. What would such a culture look like? No one can say for sure. It would be both overly simplistic and grossly insufficient to suggest: "Telecom companies must become like fast moving consumer goods companies." Simplistic, because a telecom company cannot simply transform its culture into one that stems from a completely different tradition. Insufficient, because the telecommunications sector's future will demand organizational capabilities beyond those currently possessed by any consumer goods or electronics company.

At a minimum, telecom companies must learn how to work at a much faster pace. That means, in part, evolving from today's relatively regimented organizations into more free-flowing entities that support concurrence of engineering, project management, sales, and service. Carrier cultures will also need to expand their risk tolerance, more readily accept constructive mistakes, and resist the impulse to reject breakthrough thinking as impractical and unrealistic. The most adaptive company cultures will balance their traditional emphasis on execution with the new imperatives of fostering speed and creativity, nurturing the many different kinds of talent required to actualize convergence, and creating the cohesion to make it all work together.

This is not to suggest that the transition from a culture primarily focused on operating efficiency to one that is also nimbly adaptive, innovative and smart is simple or easy. Avoiding, or at least surviving, the potential death valley between two such markedly different cultures requires vision, courage, and a sustained focus on results – as the company must continue to perform throughout the transition.

In addition to grappling with the many new strategic implications, telecom executives must learn to be increasingly tolerant of people who think and do things differently. Most leaders will be challenged, sooner or later, to reengineer their organizations to reward more risk taking and innovation, more fully develop internal talent – by exposing them, for example, to more parts of the disruption, creation and innovation process – and increasingly empower people to pursue disruptive innovation. That means, in part, leading by values versus control – that is, building the "right" behaviors into the company's DNA, then stepping back to give creative people the freedom they need to be happy and effective in their work.

Telcos have always had a network operator mindset and have been understandably proud of the state-of-the-art communications infrastructure they've built. But the days of simply providing a transport pipe have long passed. No matter where your company falls on the spectrum of players, the age of the consumer and its new terms of competition present a stark choice: transformation or slow death. Visionary leaders in the industry understand that their mix and level of talent will be the single most important determinant of success. Networks do matter. Products matter. Services matter. But ultimately, with social media, mobility, and the cloud becoming more integrated than ever, what will matter most is the talent to make new business models work.

In short, telecom executives must grow as leaders. Serious leadership growth begins with awareness that the status quo is less than ideal in the present and ma y prove untenable in the future. Both are now true for much of the sector. Forward-thinking telecom leadership teams will view themselves with fresh eyes, and will be both curious and open minded in weighing the practical implications and ramifications of convergence and digitisation.

The challenges of the new digital age make the role of a CDO a no-brainer. After all, who else will ensure that the company's key strategies are supported by emerging technologies and new organizational competencies? It turns out that many companies hire the CDO before completely defining the role. The CDO needs to excel at six key functions in order to be truly effective:

- Map digital capabilities to strategic priorities. Digital won't get the visibility or support it needs if it's not helping the company get where it wants to go.
- Serve as the executive sponsor for digital process innovation. Other executives might not be measured on new functionality or economies of scale that digital capabilities enable. For instance, consider the complexities of digitizing the mortgage loan application, and how that will impact a variety of existing business processes.
- Develop and administer the digital project portfolio. Responsibility for actual deployment of digital solutions might rest with line of business executives. Centralized monitoring of often-heterogeneous digital initiatives is critical.
- Measure new efficiencies and ROI. Replacing manual processes with technology drives both cost savings and revenue generation. The CDO's oversight across digital initiatives positions him or her uniquely well to measure success.
- Develop ways to attract and retain top talent. Companies often have more mature visions for digital than they do skills.
- Be the "executive intermediary" during delivery. This often means enlisting involvement from other executives, re-establishing priorities, acting as a coach, or serving as tiebreaker during the rollout of digital capabilities.

Digital isn't merely a thing—it's a new way of doing things. Many companies are focused on developing a digital strategy when they should instead focus on integrating digital into *all* aspects of the business, from channels and processes and data to the operating model, incentives, and culture.While most companies say they know their customers, CDOs must make it a driving passion and core competency of the organization. With technology and customer habits changing so quickly, developing a deep and detailed view of customer behavior across all channels provides a common reference point in any business discussion and arms the CDO to challenge the status quo and make changes. For example, one CDO used the concept of customer journeys and big data mapping of these paths to show her peers where opportunities and pain points existed—and, in doing so, destroyed several myths.

Digital capabilities ultimately provide an important foundation for improving the customer experience. It's up to the CDO to identify those functions where digital is critical: for example, investing in automation capabilities to rapidly respond to customer interactions, developing sophisticated reporting and analytics capabilities to interpret customer needs, building innovative interfaces to gather customer data (for example, an alternative payment method), and creating mechanisms to deliver content and offers across all relevant channels.

CDOs can build strong foundations for change by creating a "spirit of digital" throughout the organization. That could include setting up coding days for the board or holding company-wide hackathons—one company we know even had drones flying around the atrium of its headquarters. Core to building this spirit, however, is increasing the "metabolic rate" of the organization. That starts with changing basic habits, such as having strategy leadership meetings weekly or even biweekly to help ingrain the idea of moving at a faster pace. CDOs must look at how the organization operates and find ways to inject speed into processes.

In some cases, it could be as straightforward as working with IT to automate existing development processes. But in others, it will require radically changing how the company works, such as setting extremely aggressive goals—as few as six weeks—for getting a product to market. Some CDOs do this by setting up "digital factories," which are cross-functional groups focused on developing one product or process using a different technology, operational, or managerial methodology from the rest of the company. Embedding these factories in business units has the advantage of spreading the new culture and making the digital-factory approach the norm

Some CDOs spend as much as 50 percent of their time working with external partners to build effective working relationships that take advantage of every organization's capabilities. To help bring these outside voices into the organization, many CDOs establish advisory boards of start-up leaders or create "challenger" boards of people with digital experience and expertise to review corporate initiatives and strategies. At a more pedestrian level, they regularly invite technologists or entrepreneurs to team lunches.

Building an internal network is just as important because company systems and technologies need to be flexible enough to work with outside parties. In particular, CDOs need to work with IT leaders to develop application programming interfaces and cloud-based architecture that works with a broader ecosystem of providers.

As the digital age scrambles the traditional organizational structure, CDOs must not only launch the organization on its digital trajectory but also help it fundamentally evolve. The role requires a "bifocal" approach: achieving the near-term imperative of getting things moving quickly, while setting in place the longer-term conditions of success so the organization can compete digitally. Those CDOs that succeed will truly have earned their place in the already-crowded C-suite.

Telco Digital # 8 :
Conclusions Digital

Telco Global Connect

A business transformation is digital when it is built on a foundation of digital technology. This focus on digital technology is distinct from other potential transformation drivers such as political, social, cultural, or economic shifts. The technologies and business models that underpin digital transformation are not fixed. They vary over time and also, to some extent, by industry sector and geography. Currently, the following technologies are most significantly associated with digital business transformation:

• Analytics tools and applications, including 'big data',
• Mobile tools and applications,
• Platforms upon which to build shareable digital capabilities, like cloud solutions and app marketplaces,
• Social media tools and applications,
• The Internet of Things, including connected devices and 'smart' networks.

Together these digital technologies, often cumulatively referred to as the Internet of Everything (IoE), is having a profound effect on how organizations and industries are transforming, often as a result of new technology-enabled business models.The combination of organizational change and digital technologies, in turn, has the potential to improve performance in multiple areas. Indeed, it is a mistake to restrict the assessment of performance to a single metric.

Broadly speaking, performance improvements can be achieved in the following areas: increased revenues, improved efficiency and reduced costs, faster and more successful innovation, more effective knowledge collection, sharing and use, enhanced customer engagement and customer service, and finally sustained protection against digital disruption. These performance improvements are quantifiable because they can be measured and reported. The quantifiable nature of many digital technologies, such as connected devices, big data, and social media is a key enabler of digital business transformation.

Thanks to ubiquitous network connectivity, mobile devices, tiny and inexpensive sensors, and the proliferation of data collection tools, companies can understand their business environment more fully than ever before. Companies can gain insight into their operations by embedding sensors in their production equipment, fleets, facilities, and products. They can understand what customers are saying about them on social networks with listening platforms, and know where they are by tapping into data about mobile device usage. Tools that "scrape" the web for data, and "clean" it for analysis, can collect massive amounts of information on nearly anything—competitors, macroeconomic trends, and weather patterns. Companies can also gain granular detail about their supply-chain partners—not only whether goods are arriving on time, but also whether they are produced ethically and sustainably, or have been maintained in a "cold chain" that prevents spoilage, for example. All of these data sources flow continually in the hyperaware company and are monitored in real time. (IoT)

Here's what everyone knows about the Internet of Things again. It's going to be enormous. We'll have tens of billions of devices by the end of the decade. This is a multi-trillion dollar opportunity over the next few years. All the major players in consumer electronics, mobile, cloud, factory automation, enterprise IT and more will be fiercely competing for a piece of that pie.

All this information shouts: if your company doesn't have an IoT strategy you might as well quit. According to Vision Mobile , here's an uncontroversial, but often forgotten truth. The value of IoT does not come from the technology but something else.Value is created in IoT by making sense of data, turning it into knowledge and meaningful action. It's not the parking sensor that matters, but finding a free parking spot quickly and without frustration.The Internet of Things (IoT) includes wearables, connected cars and connected objects in the home.

The number of smart connected homes could hit up to 700 million homes by 2020, rising from somewhere between 100 million and 200 million homes now, according to Gartner. But others are not so sure. In fact, the Smart Home market is struggling to move beyond early adopters into a mainstream market. Gartner itself hints at a "lack of a good business model or the immaturity of home IoT products", which "has not stopped gateway makers from trying to develop the market". Fortune magazine puts it this way: "Early adopters, venture capitalists, and entrepreneurs have bought into the idea of a smart home, but mainstream consumers haven't." And according to Argus Insights: "Early adopters have gotten what they need, and now products are not compelling typical consumers to create a connected home. Acquisitions by Google and Samsung have done little to spark consumer interest."

Consumer and enterprise technology are increasingly converging in most industries. The smart home of today will become the smart office of tomorrow, as smart locks turn into access control and smart TVs into meeting room equipment. The equivalent of wearable-sensor-driven health apps in the enterprise are people analytics, such as the Humanyze platform. And Jeff Immelt, head of GE, famously said this about data technologies developed at Amazon, Google, or Facebook making their way into the industrial world: "If you went to bed last night as an industrial company, you're going to wake up this morning as a software and analytics company." Consumer, and not enterprise technology will be the foundation for the converged future. Why? Consumer markets offer much faster product evolution and validation with customers. Consumer-grade ease of installation coupled with enterprise-grade security will be the future.

With IoT, washing machine makers can now not just deliver detergent just in time by knowing when your supplies run out. They can also recommend the right detergent, based on your usage, type of clothes, on demand. Car makers can recommend where you buy your gas, by understanding your drive journey, availability of gas stations, pricing on-demand discounts, and gas station commission. Watchmakers can command a commission from health insurers, as they can monitor your heart rate, temperature, fitness habits and determine what risk zone you are in. In short, IoT makers can now afford a negative BOM (bill of materials) "a la Dell", by subsidising the cost of hardware with the revenues from bundled e-commerce services.

Internet of Things will allow any connected "thing" to become an affiliate for e-Commerce goods that are consumed together with the "thing" – what in economics are termed complements. Any connected object could become a distribution surface and customer acquisition channel for e-Commerce goods and services of every kind and description. IoT extends e-Commerce affiliate and user acquisition schemes beyond websites, mobile and apps, into every physical object. So what should any self respecting Telco do about IoT ? In simple words GET INVOLVED or get left out !!

The challenge for operators is to find a business model that delivers value for customers and is profitable. A significant part of their challenge is determining what an IoT network architecture and business operation should look like. These dilemmas need to be resolved quickly because, within the next five years, serving the IoT market will become the critical mission for any communications service providers. As industries such as automotive, utilities, transport and logistics feel the competitive pressure of IoT, the scramble for partners to help them will accelerate. One of the principle capabilities these companies will seek of their partners will be their ability to deliver complex solutions quickly.

The prevalent business models in the Internet of Things are moving from product sales to recurring revenue, and from products to services. Industrial IoT technology creates opportunities for vendors to sell access to assets like jet engines or locomotives as a service, rather than selling the machines themselves. In the home, smart appliances (e.g. washing machines) are becoming an e-commerce point of sale for consumables (e.g. washing powder). Companies like Nest, Oscar Health Insurance, or Automatic have paved the way for moving from a 'consumer pays' model to a 'consumer gets paid' model, subsidizing devices with other revenue streams like insurance or energy company rebates.

As stated emphatically by Vision Mobile Analysis , independent Developers will become the center of commercial strategy. If you're not into developers, you're not doing it right. If you believe that the Internet of Things is only about making new, stand-alone devices and solutions, then think again. More and more key players in every IoT market build their strategy around developers who can extend the product beyond what it was when it left the factory. From Amazon and SmartThings in the home, Apple and Pebble on your watch, and Ford and Automatic in your car, all the way to ThingWorx and IBM in industrial settings, 3D Robotics and DJI in drones, and Oculus and Microsoft on your virtual reality headset – developers are key to success in the marketplace.

Lets get back to the basics : Once the motivation for transformation has been clarified (the answer to the why transform question), the next stage of the journey begins – what to transform? Digital business transformation can take many forms and smart transformation requires prioritization. There are 7 distinct categories, any of which could be transformed digitally. The categories are: the business model (how a company makes money), the structure (how a company is organized), the people (who works for a company), the processes (how a company does things), the IT capability (how information is managed), the offerings (what products and services a company offers), and the engagement model (how a company engages with its customers and other stakeholders). These categories make up the most important elements of an organizational value chain as it relates to digital transformation.

Understanding the need to transform and having a good grasp of what must change are important, but the key to success lies in the implementation. How to transform is where most organizations fail. While there is no one-size-fits-all roadmap for organizations across every sector, we found that a digital business agility capability is positively and significantly linked to both positive financial performance and the ability to respond to digital disruption. Digital business agility is composed of three sub-capabilities: hyperawareness, or the ability to detect relevant trends, information decision making, or the ability to make evidence-based decisions quickly and collaboratively, and fast execution, or the ability to quickly translate decisions into action. Digital business transformation is by no means easy to achieve; but for many firms, it is a competitive necessity.

The city of Barcelona provides an interesting example of hyperawareness .The city decided to install thousands of connected devices across the city to monitor what was going on, to improve the lives of its citizens and tourists, and to save money. For example, city planners found that one third of the traffic in the center of Barcelona consisted of drivers looking for a parking spot. Therefore, they placed sensors in parking spots and linked them to apps, so that drivers could quickly and easily find the closest place to park. They also placed sensors in garbage bins to detect the fullness level, and dynamically schedule pickups accordingly.

To win in the Digital Era, companies must be dynamic enough to understand the risks disruptors pose to their business, and to maximize their revenues in the face of tough competition. At the same time, the percentage of revenue from new businesses needs to increase, as margins from core businesses will frequently decline. This puts pressure on companies to improve the "hit rate" of marketable innovation, an area in which nearly all incumbents struggle.

Companies must make decisions based on insights gleaned from data analysis, and ensure that experts from within and outside of the organization have access to these insights. Experts must be brought into the decision-making process at appropriate stages, regardless of their location, role, or rank. This requires an inclusive business environment in which experts have the opportunity to participate, can provide contrary evidence, and give frank recommendations, even when they contradict the opinions of senior leaders. Diversity of perspective (from contributors of different genders, races, and creeds) contributes to informed decision-making but is only one element of a larger imperative of corporate inclusion. (Global Center for Digital Business Transformation IMD Lausanne)

Claudio Contini, CEO of Telecom Italia Digital Solutions, says: "Customer experience and user numbers are the new sources of revenue. More than ever, today, the customer is king. Only by understanding and embracing the need to fundamentally change the way we do business can we remain relevant to our customers." "The first steps are essential. Planning requires a review of the company's strategic goals, assessment of the current environment and a roadmap based on impacted areas such as people, process and tools - which, combined correctly, will help steer the transformation journey in the right direction."

An operating model that is not prepared to handle the strong shift to digital communication leaves service providers vulnerable to missing out on opportunities to improve the customer experience, streamline their business processes, and ultimately stay relevant to customers.The transformation to a Digital Service Provider represents great opportunities for operators to drive innovation as they develop strategies to communicate with and deliver services to customers across the all platforms.

According to GSMA "Beyond connectivity, mobile operators will play a crucial role in working together with a range of industry partners in health, automotive, education, smart cities and a range of vertical industries to accelerate the launch of valuable connected services," Unfortunately without continued investment and growth in mobile networks (especially LTE) and the deployment of multiple connected devices, the socioeconomic benefits of the connected life will not be fulfilled.

mHealth programmes are currently one of the most cost-effective ways of providing remote living assistance to aging and chronically ill patients. mHealth programmes provide faster response times, integrated record access and considerable ease of use to patients. Remote consultation and support is expected to address the growing chronic disease management issue by reducing the need for hospitalisation. Proactive mobile based care for patients with sudden health incidents can reduce the number of primary and emergency visits by 10%. Mobile technology can also be used for home monitoring, thereby reducing the need for face-to-face consultations.

At the Barcelona MWC , Telcos had the opportunity to generate value beyond basic connectivity through managed connectivity, stewardship services and platform innovation. The GSMA area was filled with interactive demonstrations of the connected life including the Aston Martin One-77, the bike of the future. It is fully connected and tuned into its own performance as well as the rider, including mobile health monitoring and electronics that track the bike's performance in relation to its environment.

There was also the Mantarobot showing virtual teaching through augmented reality and virtual presence and the Cooltra Connected Electric Scooter, the latest in smart city transportation, the GO! S3.4 from GOVECS which lets customers know when and where scooters are available and can be started with a phone via NFC. The widespread penetration of mobile networks offers a powerful platform to improve access to relevant content.

mEducation solutions already allow thousands of students in China, Bangladesh, South Korea and Indonesia to access course content through SMS and audio lessons. An mLearning student saves 86.7% of the cost spent by students taking the same training in a traditional classroom. Much of this is due to the elimination of the cost and inconvenience of travelling to attend courses. Inexpensive personal learning devices like the 35 USD tablet launched in India are further improving access to mEducation.

In developed countries mobile interventions could help cut healthcare costs by 400 billion USD in 2017, help retain 1.8 million students in the education system, save one in nine lives lost in road accidents, and reduce CO2 emissions by 27 million tonnes annually. Similarly in developing markets, mobile interventions could help save over a million lives in Sub-Saharan Africa, provide education access to 180 million students, save 25 million tonnes of food and encourage over 20 million commuters to start using public transport. (GSMA Connected Living Program)

Mobile networks play a pivotal role in the development of the connected life providing a scalable, standardised global platform to support the growing demand for intelligent, secure connectivity.Examples of valuable connected services were amply demonstrated by leading edge Telcos at the GSMA Connected City showcase in Barcelona included :

In the Connected Home AT&T showed how people can use their smartphones and tablets to manage their energy, automate appliances and secure their homes through AT&T Digital Life. General Motors demonstrated how AT&T's 4G LTE network will transform the driving experience by enhancing safety, security, diagnostic and infotainment in the vehicles starting next year.

Deutsche Telekom, in conjunction with IBM, bought to life Smarter Cities for the Future using machine-2-machine technology to optimise urban services such as public transport, parking, energy, security and water management. Together with SAP, Deutsche Telekom also showcased Connected Port Solutions designed to optimise both road and sea traffic control as well as logistics and terminal operations in order to make port processes more efficient allowing larger quantities of goods to be trans-shipped in the port area.

Korea Telecom featured technologies that make our lives better including edutainment robots, automatic content recognition, smart home phones, a controlled motorcycle, eco food bins and cloud CCTV.There were also Smart Apps in your hand showing how we can live smarter with intelligent and unique applications including mobile K-pop music, integrated mobile payment and self-created M-learning solution.

Vodafone were showcasing their Smart Home, Smart City and Smart Mobility solutions.Smart Home illustrated how M2M technology can provide premium security services, enable remote health monitoring and even open and close doors remotely.Smart City demonstrated how Vodafone's Energy Data Management (EDM) solution, solar energy production monitoring, remotely controlled street lighting and digital signage are enabling the smart city. Vodafone's Connected Cabinet solution demonstrated retail display cabinets that report on location, operational status and stock levels in real-time. Smart Mobility showed how M2M is transforming the automotive and transportation industries, be it through real-time information systems for public transport, enhanced drivers' experience with telematics services or through usage based insurance services with Vodafone Vehicle Connect.

So lets summarise quick : Going Digital means retooling the network for Virtualisation , Cloud , CXM, reconfiguring the Organisation structure , revamping the IT Core with analytics , creating a new business model based on thorough understanding of the digital consumer....YES EVERYTHING AND MORE. So if your CxO team lack the intestinal fortitude for a difficult transformational journey ...my sympathies :)

POST SCRIPTUM : Who will be the Digital Titan ?

Do you really want to know who will dominate the Telco space in the next decade ? Do you really wanna be a Global Telco Titan ? If you do then you might want to track the activities of the Data Centre Titans . Every quarter, cloud giants Amazon, Microsoft, IBM, and Google collectively spend billions of dollars on servers and other hardware for their cloud services and data centers around the world to house all that gear. Cloud service providers consistently say that data center spend represents the biggest portion of their capital expenditures. Microsoft, for example, spent $1.5 billion in the last quarter to support demand growth for its cloud services. Amazon has 11 regions around the globe where it has datacenter capacity.

Amazon scales up the capacity in each availability zone by adding whole new datacenters. AWS designed its own network routers and went to original design manufacturers to build the hardware, and put together a team to write the networking software stack on top of them. Today all of the AWS network is using this custom network stack. And Amazon has a HYPERSCALE ambition that should scare some Vendors and SP's out of their wits !!

Now what you my ask is all this Data Centre expansion to do with Telco industry ? How about everything !! To get some ideas you might want a peek into Bell Labs Future X Network Vision. It talks about the new cloud-integrated network (CIN) which is comprised of two related and essential architectural elements: the edge cloud that is embedded in the network; and a re-partitioned access and aggregation network that delivers ultra-high bandwidth and ultra-low latency to and from this edge cloud. In turn, this distributed cloud connects to the centralized cloud that hosts less performance-sensitive applications, over a similarly adaptive core network.

The global competitive landscape will change dramatically because of Cloud and Data Centers. Expect the emergence of a new global-local duality of new service providers with either global or local focus like maybe approximately 10 global service providers, that offer global connectivity, cloud and contextualized control and content (C&C) services. Then a set of approximately 100 local cloud-integrated network providers that offer domestic, hyper-local connectivity, edge cloud and contextual C&C services and the formation of a global-local alliance framework that connects and interworks the local and global service providers, similar to that of the airline industry.

We will see the emergence of new virtualized enterprises, which are completely location-independent, and enjoy automated connectivity to and between any process or person or thing, with the ability to scale to meet any demand. And how about the rise of new digital commodity exchanges that redefine existing businesses, by leveraging the global-local cloud infrastructure and dynamic connectivity to allow the facile exchange of digital goods or digitally-connected physical goods. This new innovation continuum, which Bell Labs call continuous disruptive innovation, in which the technological separation between sustaining innovation (by incumbents) and disruptive innovation (by new entrants) disappears.

You can bet your last penny that the Web Giants like Amazon have factored in the above assumptions into their Data Centre expando strategies. They know that a massive amount of Data will be gathered because of IoT , Analytics and stuff. Data needs a safe and secure home right ? Which brings us to the role of Telcos in the new world of Cloud and Big Data. We already know that breakthrough technology advancements of the recent years (e.g., SDN, NFV, big data, All-IP) will change the way networks are being constructed and managed. These changes will enable the development of a highly flexible infrastructure that allows cost-efficient development of networks and associated services as well as increased pace of innovation.

One of the operator's key assets is infrastructure.Infrastructure usually is used by an operator to deliver own services to the end-customer. However, especially in the wholesale business it is common that parts of the infrastructure – so-called assets - can be used by a third party provider. Assets can be different parts of a network infrastructure that are operated for or on behalf of third parties resulting in a service proposition. Accordingly, one can distinguish between Infrastructure as a Service (IaaS), Network as a Service (NaaS) or Platform as a Service (PaaS). Another dimension of asset provisioning is real-time network sharing that refers to an operator's ability to integrate 3rd party networks in the MNO network and vice versa, based on a dynamic and context dependent policies (e.g., congestion/excess capacity policies).

The big question is : Should the Telcos invest more in connectivity infrastructure (4G , 5G , Ftth etc) or the data centres (EPC, SON , Cloud).If the likes of Amazon , Rackspace , Google are already throwing in billions into data centre capacity would it be more economical to collaborate with them ? Like it or not you will require massive data centres all over the world especially when the cloud and software is eating up the network. Well Deutsche Telekom (DT) does not want to collaborate because last week their Open Telkom Cloud went live. Deutsche Telekom claim that its cloud will be a much cheaper platform than that of Amazon Web Services !! Aaah yes...your data is secure fellas so let us house it for you real private No danger of Yanks snooping into your data.

That is a brave and proactive move from a Telco (DT) that wishes to be a Titan not a pygmy in the next decade . Its called strategic positioning in a world that will be swamped by petabytes of data.Unfortunately not every Telco can achieve this since their capex priority is connectivity infrastructure and data centre is just an ancillary cost to deliver the content / apps and manage the connectivity network. But then ....not every Telco CEO has a Global Vision at Hyperscale a la Google , Amazon , Facebook...or the slightest clue about what is about to happen especially when 5G blasts us into light speed...well some of us anyways !!

By the way DT is one company that has made the Open Source Cloud as the centrepiece of their global services strategy. By employing the cloud service brokerage strategy, Deutsche Telekom has become the cloud partner that its millions of business customers are looking for.Deutsche Telekom enhances its set of cloud technologies with OpenStack. The open source cloud operating system makes it easy for software partners (ISVs) to integrate their cloud applications in the Deutsche Telekom infrastructure and its new Business Marketplace.

DT's aspiration is for €2.5bn in additional revenue by 2016 built on a track record of being among the first to bring leading-edge cloud innovation to customers in Europe covering domains like end user computing, enterprise networking and data center. DT is not afraid to innovate on its own behalf; indeed, it develops many of its applications and platforms in-house. It is, however, become very open to working with partners if they can either speed up time to market or add brand credibility.

This futuristic vision of light speed connectivity supported by the Cloud and Data Centres would, on the surface, seem to be the stuff of science fiction, rather than rational technological thinking but we are at a tipping point in time. Its all about Outliers and Black Swans now. Meaning what you think does not exist and some competitor out of the Cloud can blow you out of existence if you are careless.

The so called CIN (Cloud Integrated Network) will blur the lines between application, device and network functionality by linking together device software functions with application functions and network functions in dynamic service chains, on demand. The Telco Industry's 5G network with seemingly infinite capacity will be created, and it will blur the line between the cloud and the network, as the cloud becomes distributed within the network for the required low latency performance.

Visionary leaders in the industry understand that their mix and level of talent will be the single most important determinant of success. Networks do matter. Products matter. Services matter. But ultimately, with social media, mobility, and the cloud becoming more integrated than ever, what will matter most is the talent to make new business models work. So if you " get it " you might succeed as a Master of the Game...and if you dont't..well you can still aspire to pick up business just like those opporturnistic birds that fly into the mouth of a basking crocodile and feeds on scraps of food and leeches attached to the jawline and tongue...Guten Appetit !!

FOR DESSERT HERE IS MY CHOICE OF THE BEST quotes that captures the spirit and content of the Digital Era expressed at the Mobile World Congress in Barcelona !!

"Everything that can be digitized will be digitized. Everything that can be connected will be connected. Bridging the infrastructure and ecosystem to digitization is a mega trend we see in the second phase of the Internet "

(Tim Hottages , CEO Deutche Telecom)

"The industry focus in the coming years will be on personal data, connected living, digital commerce and networks. To realise the necessary developments in these areas, "huge investments of $1.7 trillion need to be done from now until 2020",

(Jon Fredrik Baksaas,CEO of Telenor and chairman of the GSMA)

"Everybody worries about being a dumb pipe, and whether revenues will be able to support network investment that we need to make.Any other industry would be excited and highly optimistic given the strong demand in growth for their core services. However, the big problem we have as an industry is we have been unable to monetise this increased demand "

(Chua Sock Koong , group CEO at SingTel)

"Communication networks are facing a lack of scalable and sustainable architecture to meet the challenges ahead in terms of data traffic increases, video uploads and downloads, and enhanced M2M communication. The network of the future has to be highly elastic in order to facilitate the adding or dropping of capacity and real-time provisioning of service. It needs to be highly orchestrated by key business imperatives, such as customer satisfaction, and it must be highly integrated so that synergies are fully embedded and captured across fixed and mobile, across borders and across segments."

(Bruno Jacobfeuerborn Deutsche Telekom CTO)

"We have a very precise directive to establish asymmetric regulation where a company holds more than 50 per cent market share.This concentration of telecoms services to a few operators has resulted in uncompetitive pricing and obstructed new entrants. 70 million Mexicans have no access to broadband, while those that are connected experience low data speeds and poor quality. We need to leapfrog this situation."

(Mexico Regulatory commissioner Adriana Labardini)

"Why are the next two billion not on the internet?. The reason is not because they don't have any money, it's because they don't know the value of having a data plan or the services they can access."

(Mark Zuckerberg, founder and CEO of Facebook)

Regions around the world have always had initiatives regarding new technology, and it's important that you start early with R&D to stay in the forefront. For example, we first commenced work on 4G/LTE in 2000, and understand the need to invest in the research of 5G."

(Ericsson's CEO Hans Vestberg)

"To change the vicious circle to a virtuous one, we're going to need more cooperation than there has been in the past. Then we're probably going to get to a win, win, win solution, as opposed to when you used to have people duking it out [as to who 'owns' the customer] in the past.There's greater value in cooperating than just competing. Developers cando this too by opening up their APIs"

(Doug Webster, CMO, service provider, for Cisco)

"While big data is great, there is a lot of it out there in silos, and each data [set] speaks its own language, you need to be able to solve issues around homogenising it, then you need to solve issues around analysing the data... what we need is a 'universal translator.We also need a better policy over who owns this data. We need to have joined up thinking about these open platforms,"

(Young Sohn, Samsung Electronics, president and CSO)

" NFV and SDN concepts are at the core of our strategy. These help us realise our future network vision, which is a mutli-service, multi-tenant platform where we can respond more quickly and efficiently to our customers' needs. With NFV, we're able to dynamically reroute traffic and add capacity without adding new boxes. With SDN, we're removing pre-defined physical limits of the network by shifting control from hardware to software. These allow the network to become simpler, more scalable. They also allow us to reduce costs significantly and more quickly address customer needs "

(John Donovan, Senior VP– Technology and Network Operations, AT&T)

"Not everything can be automated at once and municipalities need to select carefully the value added services that will give consumers the most value for their initial offerings. Once a base is established the municipality can market the platform as an opportunity for other businesses that wish to deliver value added services. "

(Jeff Edlund, CTO, CMS, Enterprise Services, HP)

"Between 15 per cent and 20 per cent of subscribers say they have poor coverage at home, and yet more than half of homes in countries like the USA and the UK have Wi-Fi. Wi-Fi is an obvious choice for solving the problem."

(Ken Kolderup, Kineto's CMO)

"We're attracting new vendors by offering them an alternative to a mobile ecosystem that depends on two restricted operating systems with many strings attached. Apple controls all aspects of their offering and eschews customization. Even on Android, from an OEM perspective, only one manufacturer is profiting. Many people in the industry want to see a third option become more viable, and our flexibility makes us an ideal partner."

(Jay Sullivan, COO Mozilla)

" Mobile connectivity hold huge potential for women – currently an untapped economic potential. The Internet is proving to be an effective catalyst in transforming gender opinions. Access to a communication network provides women with the flexibility of working both remotely and on their own terms, allowing them to build independent companies. The more we can work together to develop this offering, the greater opportunities we can build for the region's women as a whole"

(Dr Nasser Marafih, Group CEO, Ooredoo)

"Our customers trust us, so our responsibility of how we manage this private information will become higher. Operators will have to become involved in defining how this sort of data is used, while protecting the privacy of the individual."

(Kaoru Kato, Docomo's president and CEO)

"Operators are] shrinking in relevance too fast. Many partnerships (with content and internet players) have been for loyalty, not for revenue streams. We need to find different ways of charging our customers rather than complain "

(Johan Dennelind TeliaSonera CEO)

"We will look back on this time and look at data as a natural resource that powered the 21st century,"

(IBM CEO Ginni Rometty)

" The mobile industry should consider 5G as a "special generation", introducing challenges in all layers of the technology"

(Mike Short, VP of public affairs Telefónica Europe)

" We have engineering teams working on LTE and 5G. Each time the 5G team unveil a new performance leap, the LTE engineers respond by matching it "

(Matt Grob, VP & CTO for Qualcomm)

"We really do need 5G in order to have a paradigm shift. The order of magnitude jump in traffic is what is really driving this move."

(Mischa Dohler, professor of wireless communications at King's College London)

"5G is a fundamental change in technology and will have a significant impact on how we offer services. We must look at performance and coverage, and not just consider microcells."

(Allan Kock, director of RAN development at TeliaSonera)

" 5G must have green as part of its very DNA, ensuring that all aspects of 5G, from access networks, data centres and transport network to connected devices only consume energy when they are being used. However 4G is a success, let's enjoy . We shouldn't jump too fast," "

(Stephane Richard Orange France)

"Most of our competitors talk about 5Gb/s and 10Gb/s or some other number, but they're not telling you the configuration, which is ridiculous. 5G should not be a "forced leap" in technology, but draw heavily on re-use existing radio air interfaces – LTE, WiFi and LTE-U – under a common control plane."

(Marcus Weldon, CTO of Alcatel-Lucent)

"Computing costs have fallen 1000-fold since their inception.We've achieved this change with semiconductors, we now have to do the same with 5G compared to 2G,"

(KT Telecom CEO Chang-Gyu Hwang)

Ultimately, the success of 5G will depend on the success of the entire ecosystem, one in which innovation will become the key driver behind 5G development market demand.In the GSMA hyper-connected vision mobile operators would create a blend of pre-existing technologies covering 2G, 3G, 4G, Wi-fi and others to allow higher coverage and availability, and higher network density in terms of cells and devices, with the key differentiator being greater connectivity as an enabler for Machine-to-Machine (M2M) services and the Internet of Things (IoT).

Bottom Line on 5 G : Hold your breath till 2020 if you insist on 5G or focus on monetising LTE and LTA A now !!

" There is a clash of business models between network operators and internet firms, and that network players were left holding the short straw. We are an asset-heavy industry, where everything is interoperable and open while OTT is asset light. How can you compete with a voice, SMS or video service which costs nothing?"

(Timotheus Hoettges, Deutsche Telekom Group CEO)

" Operators are competitively disadvantaged with over-the-top players, owing to heavy infrastructure investment. Regulators should apply the principle of "same service, same rules" to players of every hue, and that consumers should have a "portable digital life" where it's just as easy to switch digital ecosystems as it is networks.

(Caesar Alierta , CEO Telefonica)

" Offering free basic internet services can help mobile operators grow their businesses faster in emerging markets. And once users get a free taste of the internet they'll be more inclined to pay for mobile data.The Ebola crisis in West Africa is an example of a place where a crisis may have been exacerbated by the lack of good connectivity," "

(Mark Zuckerberg , CEO Facebook)

"Our goal here is to drive a set of innovations (like Project Loon). I think we are at the stage when it's important to think about hardware, software and connectivity together. We are very happy to work with Facebook on Internet.org" "

(Sundar Pichai, SVP of products at Google)

"Those opposed to open internet rules, they like to say we used depression era regulation. But we took Title II and modernised it. We built our model on a regulatory model that has been wildly successful in the US for mobile."

(Tom Wheeler FCC Chairman)

"I think you almost have two choices as an operator: either you say that you can be very, very efficient with pipes and that somebody else can deal with all of that and you're going to have the lowest cost and the best quality on my pipe, transmit as much data as possible "

(Hélène Barnekow TeliaSonera CCO)

It all seems so unfair that OTT players make their money by loading more and more traffic on the operators' networks, thus causing the operator to pay for network capacity upgrades to facilitate the OTTs' increasing VoIP and IM traffic that was the root cause for decreases in the operators' core voice and SMS revenues. Telcos you can't wish the OTT's away anymore than you can do the same with your mother in law !!

Bottom Line : Collaborate or compete the consumer will have his cake and eat it too !!

"We are clearly looking for distressed assets. If we can find the right opportunity, within specific countries, we will do that, "Big operators buying up whatever they can are rarely a success case. It bites them in the behind later. Operators "should start with in market consolidation, then fixed mobile convergence ... then international – the latter is very complicated"

(Mats Granryd, CEO of Tele2)

"Authorities are starting to understand that the industry is setting itself up for a data only world. The Europen Commission is a bit of a Dr Jekyll and Mr Hyde. When you discuss the merits of a deal they are extremely conservative. When you come with convergence and OTT as an argument they are not very receptive. The commission views market consolidation as a question of] fixed cost efficiencies,"

(Thomas Wessely, partner at Freshfields)

" The value of mergers and acquisition in the technology and operator sector grew 34.9 per cent in 2014 from the previous year, according to. The last time it was that high was in 2000," said Thrasher. And last year telecom M+A as represented three of the ten biggest deals"

(B. Holt Thrasher, MD for Mooreland Partners)

" We believe M+A as a potential means to transform its business as the industry moves towards software based service delivery. Our previous approach was more focused on filling gaps. We are now looking at a complete view of which new business we need to be in. The performance on acquisition is not necessarily stellar. It's an important caveat. An acquisition is not a panacea, it's not going to solve the problem"

(Rima Qureshi, SVP, head of M+A, Ericsson)

There is no substitute for old-fashioned focus on the fundamentals of MA: a clearly articulated and well thought-out strategic rationale for the acquisition should become the yardstick by which to measure individual decisions that arise during the course of a transaction. Despite the clear economic rationale, there may still be some obstacles to consolidation, such as the unrealistic ambitions of owners and senior management or an investor reluctance to take a financial write down. While in-country MA has led to improved profit margins for the merged business, this has, in most cases, come at the price of a loss of market share

Bottom Line : Do your Due Diligence because mating two turkeys does not an eagle make !!

"The pace of change in today's mobile ecosystem sometimes makes it difficult to keep focus on core business areas like roaming that still carry tens of billions of dollars in revenue opportunity, Things (IoT) and wearables. It warned mobile operators that these newer innovations could be "market distractions" and carry an opportunity cost of up to a staggering $46 billion."

(Mary Clark, Syniverse)

" Given that M2M ARPUs are low, the cost of supporting sensors has to be significantly below that of supporting subscribers. If operators are going to cost-effectively manage a huge number of new devices in their networks then they need an understanding of the machine's environment and enough information to trigger rapid automated responses to any changes that affect service performance

(Anukool Lakhina, CEO and Founder of Guavus)

"The car of the future is a smartphone on wheels." Dieter Zetsche, Chair of Daimler AG

" We believe cyber security and approval from regulatory authorities will be key challenges when it comes to connected cars, but says both Nissan and Renault know exactly where they are headed with "autonomous" cars (not driverless).

(Carlos Ghosn CEO at Renault – Nissan Alliance "

" It says something about the achievements of the mobile industry over recent years that many IoT propositions appear to take the mobile connectivity component for granted: there is an assumption that it will simply work. Interoperability will remain a critical success factor for IoT. Interoperability still matters. Any IoT device with poor interoperability will struggle to achieve scale. "

(Lars Nielsen, General Manager, Global Certification Forum)

" From an ATT perspective, we are providing an end-to-end (E2E) platform that enables a unified experience across a wide range of devices and one-off capabilities in the market today. There's a lot of noise in the marketplace that can make it complicated for a consumer or a business to make sense of this notion of the Internet of Things. It is important for us to show real benefits, whether you're a consumer or business "

(Kevin Petersen, ATT Digital Life President)

" The SIM card is evolving to support future new services in both the machine-to machine (M2M) and consumer markets. By 2020 we forecast nearly 1 billion cellular M2M connections and 9 billion consumer connections that will require SIM cards, so it is critical to ensure security and robustness in the evolution of the SIM."

(Hyunmi Yang, chief strategy officer GSMA)

" In the emerging Internet of Things (IoT) space telecoms operators will likely continue to provide connections into consumers' homes and power the different screens. But to remain relevant within the connected home landscape , operators will need to find new ways to interact with the things people care most about in their homes – things that help them stay comfortable, help keep them safe and help them save energy.

(Chris Borros Nest Labs)

"It's (IoT and Digital Life Platform) not just a differentiator, it's an imperative for success. Nearly every CIO I talk to has security as his or her number one concern. What IoT can do for businesses is so exciting, but customers want to know their data is secure," (Ralph de la Vega, president and CEO of ATT Mobile and Business solutions)

Wearable computers, personalized advertising billboards and self driving cars are some of the items that appeared in science fiction blockbusters like the Minority Report to depict a futuristic world. Now, in 2015, this scene has become a reality. With the spread of smart devices, evolution of mobile networks and growth of smart technologies encompassing sensors, cloud computing and Big Data, the era of the Internet of Things (IoT) has finally arrived.

Bottom Line : Jump on the IoT bandwagon for the right reasons. Be prepared.. its not an trial initiative but a way of life !!

As such the best view point came from, when he said " 70 per cent of the operator's customers switch off data roaming when travelling abroad. But in the future we want to get rid of bill shock, so that mobile systems basically serve us. They deliver better peer-to-peer connectivity, better security and better privacy in the long run. Removing the bill shock phenomenon needs new pricing initiatives to create a new level of affordability".

Jon Fredrik Baksaas, Telenor Group CEO

So while Mobile Operators invest billions in the mind boggling capabilities of new technologies that will they should also try to resolve their subscribers current pain points..like BILL SHOCK. That will earn you new revenues , improve CoE , reduce churn...and less cause to complain about the OTT's eating your lunch !!

MWC 2016 Barcelona Day 1 : CEO Key Notes...Blue Skies in the Digital Firmament !!

The opening Key Notes are always inspiring in terms of "digital pie in the sky " thinking and this year is no exception. A lot of it is hot air and fluff but the Vodafone CEO in particular impressed me with his impassioned call for the Gigabit Society. In Vittorio Collao's view a Gigabit Society is where citizens and businesses benefit from widespread connectivity of 1 Gigabit per second delivered by robust, reliable and future-proof fixed and mobile technologies. The 1 Gigabit per second downstream capability should be combined with significant upload speed capability to deliver true ultrafast broadband. Countries cannot afford to invest in technologies of the past like copper and instead needs to focus on the no regret option of investment in FTTH, cable and 5G mobile networks, the technologies that can deliver ultrafast broadband.

In Collao's (and Vodafone's) desire for a Gigabit Society , only supportive policies and activities by government and relevant organisations will ensure its success. Increasing digital literacy and raising awareness of the benefits and possibilities created by forward-looking digital infrastructure will be important to ensure all citizens maximise the potential of this new and vast opportunity. That is what Cesar Alierta , CEO Telfonica , alluded to in terms of dire need for training and education in the Digital Society. Meaning structural unemployment ..because there will be jobs but no people with the requisite skills to take them. As if this isn't already a major problem in developing countries !!

In this context, the attractiveness of regions and municipalities and their long-term ability to maintain and develop economic activity will become more dependent on the digital infrastructure they offer. This is supported by recent OECD research on the effect of local

FTTH networks in Sweden which shows that a 10% increase in fibre penetration is correlated with 1.1% higher employment and greater business creation.

Increasingly, all sectors depend on customers having access to high-quality reliable infrastructure and services, including the cloud, government services and digitally provided education. Capital and labour are increasingly mobile with some service industries becoming super-mobile (e.g. start-ups, programmers, creative industries) and highly dependent on ultrafast broadband infrastructure. Take a look at the how the Dutch built up their ICT Infrastructure :

Virtual reality came up as a topic by all the opening speakers. While all of this sounds exciting, the additional strain that VR will put on networks is a major concern. To be able to support data-powered innovations such as VR, global connectivity providers have invested heavily in capacity, engineering a future-proofed backbone that shouldn't buckle under the extra pressure. The biggest challenge lies in last-mile networks which need to be able to carry VR content to homes and provide a seamless, genuinely immersive experience for consumers.

BT has identified a number of VR applications it believes will change the world. The first, unsurprisingly, is gaming, and the second is film & television. The telco also believes that virtual travel will also become At the BBC, home of iPlayer, one of the most successful on-demand Internet catch-up TV and radio services in the world, Matthew Postgate, the broadcaster's Chief Technology Officer, says the organization is taking "a new structure and approach" to be able to respond to how the Internet is changing the distribution and consumption of broadcast TV, including a future he believes may well be characterized by immersive media. Key VR applications, include virtual voyages into space, while also expecting virtual education and remote surgery to drive social change

VR requires about five times as much bandwidth as HDTV, as well as very low latency to support an immersive experience. While large service providers offer the backbone capabilities, cloud footprint, advanced traffic management and content delivery networks at the core, many of today's access technologies at home do not support such requirements, preventing high quality live streaming of VR content from the cloud. Instead, the user will need to buffer or store the content locally, which greatly limits commercial opportunities with VR, starting with online gaming. This means intelligent traffic management solutions, compression algorithms and investments in very low-latency, high-throughput networks will be necessary for last-mile networks to cope with the demands of VR content.

However, a major question remains – can telcos find a way to monetize these immersive applications over & above the revenue they receive for distributing digital content to and from subscribers? And by the way this is the greatest challenge for Telcos in the next decade : HOW TO MONETISE technology investments in the Digital Era !!

Sadiq Malik (Telco Strategist)

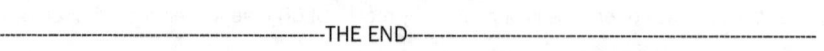

--THE END--

www.ingramcontent.com/pod-product-compliance
Lightning Source LLC
Chambersburg PA
CBHW070259190526
45169CB00001B/476